ALSO BY LARRY KILHAM

Nonfiction
Great Idea to a Great Company: Making Inventions Pay
MegaMinds: Creativity and Invention
Winter of the Genomes
The Digital Rabbit Hole
Shades of Truth / Los matices de la verdad
Adventure Skiing in the '60s
The Perfectionist: Peter Kilham and the Birds

Fiction, based on AI
Love Byte
A Viral Affair: Surviving the Pandemic
Saving Juno
The Juno Trilogy
Free Will Odyssey

Destiny Strikes Twice

James L. Breese

Aviator and Inventor

Larry Kilham

futurebooks.info

Destiny Strikes Twice: James L. Breese Aviator and Inventor
Copyright © 2020 Larry Kilham

All rights reserved. Except as permitted under the U.S. Copyright Act of 1976, no part of this publication may be reproduced, distributed, or transmitted in any form or by any means, or stored in a database or retrieval system, without the written permission of the author except in the case of brief quotations embodied in critical articles and reviews.

Published by larrykilham.net v1.6
Photo credits: Front cover, U.S. Navy, back cover, Mari Angulo, Artotems Co. All other photos from the Breese family collection unless otherwise credited.

Available for purchase from Amazon.com.

James L. Breese, Jr., USA, 1885-1959
Biography – Adventurer, Engineer, and Entrepreneur

ISBN: 9798569120215
Independently published

For my mother
Frances Breese Forbes
who wanted this story told

Contents

Preface

1 – Growing up..1
2 – Preparations for the Transatlantic Flight..... 9
3 - The First Flight Across the Atlantic............19
4 – The Chicago Interlude........................... 49
5 – Jim Discovers Santa Fe..........................57
6 – The Fun Thirties.....................................64
7 – Building the Oil Burner Business...............76
8 – Partners and Successors..........................86
9 – Troubles with Patents............................ 93
10 – The *Loke*... 99
11 – Florence.. 108
12 – The Last Flight.....................................114

Epilogue..121
Acknowledgments.......................................123
References.. 124

Preface

The time is overdue to tell the story of my maternal grandfather James L. Breese, an amazing technical entrepreneur of the Age of Invention. Jim Breese was famous for being the flight engineer on the first flight across the Atlantic in 1919. He moved to Santa Fe, New Mexico in 1929, and from then to 1959 he built an oil burner business with a portfolio of over 130 patents. All the people who knew him well have passed on. Fortunately, I knew many of them and they gave me a lot of insights and reference materials.

I believe I can understand Jim Breese because our lives had some striking parallels. Our families were upper-middle-class but none managed to hold on to great wealth. Jim and I were both very adventurous in our youth before we settled down to build businesses based on our inventions. I knew Jim personally because we lived much of our lives in Santa Fe, New Mexico.

I hope this book adds helpfully to the literature about inventors—who they are, why they came to be, and how they succeeded. But I'm afraid the analysis is never complete. Inventors are like painters or sculptors or artists: they are driven by an inner spirit that reveals a truth beyond ordinary vision.

Besides creating a historical document about a major aviator and inventor, I hope to provide an engaging story to inspire and guide the new generations of innovators.

Destiny Strikes Twice

1

Growing Up

Many years later, as he sat under his favorite willow tree contemplating all his problems, Jim Breese thought back to the time when his enormous aircraft seemed unlikely to take off on the first flight across the Atlantic. The engines wouldn't start. He had one option left: to increase the ignition voltage—a non-standard procedure. The engines had an eight-volt ignition system and he would try connecting a 12-volt standby battery. He pushed the battery into position and connected it to the ignition circuit by bridging with his open jackknife. On May 16, 1919, the four engines of the Navy NC-4 coughed to life and the huge flying boat taxied for take-off on its history-making flight. Problems were fun then. All of life was a great challenge.

My grandfather, James L. Breese, Jr. became a prodigious and adventurous inventor. Those were the days. Thomas Edison and Alexander Graham Bell were still working. World fairs were showing all kinds of marvels of technology and engineering. Artists and architects were producing monumental works. His story should start with his remarkable and creative family.

Jim, as he preferred to be called, was born in Newport, Rhode Island in 1885. He and his two brothers and sister grew up in Southampton, Long Island, starting in 1898 in "The Orchard," a mansion renovated for his father, James L. Breese, Sr. His architect was his close friend, the legendary Stanford White, who preserved its stately white Greek Revival style. The house and grounds would have been the envy of the Great Gatsby.

Jim's boyhood home, *The Orchard*

Jim's father was the nephew of Samuel Finley Breese Morse, the inventor of the telegraph and a noted painter. James L. Breese, Sr. studied engineering at Rensselaer Polytechnic Institute. He was known for his lavish and risqué parties, as a patron of the early race car designers and drivers, and as an investor in an early aircraft venture. He was also a talented amateur photographer. He and other well-known photographers called "The Carbonites" used a difficult printing process in his Carbon Studio. To finance all of this, he was a Wall Street investor who made and lost millions. Between 1909 and 1916, he made $2 million ($50 million in 2020 dollars).

Jim's mother, Frances Tileston (Potter) Breese, was the sobering influence. She was the daughter of a Civil War Army general, Robert Brown Potter, and the granddaughter of an Episcopal bishop. Her daughter, Jim's sister Frances, wrote about her mother: "After the marriage, she suffered emotionally from trying to reconcile a strict religious upbringing with the pace set by an active, irreligious husband whose lifestyle was more innovative and bohemian than hers...She could never have uttered the word sex, or, I imagine, enjoyed it." She died in 1917 at the age of fifty-three.

You might think an imaginative youngster would visualize himself as the lifelong owner of this estate, or at least spend his years pursuing New York and Long Island's social pleasures. But Jim Breese's mind was roaming elsewhere. He was fascinated by mechanical things and he sought achievement through invention. In 1909 Jim started a lifelong habit of keeping a pocket invention notebook to jot down ideas as they occurred to him.

It all began with his father's passion for the new sport and business of racing cars. In those early and simpler times, new designs were often produced even in garage shops. Jim Breese, Sr., was a governor of the Automobile Club of America, a member of the Automobile Club of

France, and a member of the Racing Board of the American daring sport, in 1904 he drove in the Long Island

James L. Breese, Sr. as a young man

Frances Tileston (Potter) Breese

Vanderbilt Cup Race, The Daytona-Ormond Beach Race, and the Edison, New Jersey, Eagle Rock Hill Climb. He enthusiastically introduced his three sons to the emerging world of automobile design and construction.

Jim Breese, Jr. was sent to Groton, an exclusive New England Episcopal boarding school. That choice was fortunate because some of his classmates turned out to be of great assistance in his business career. He graduated in 1905 and then went on to Princeton, and in 1909 he received a degree in civil engineering. There is no indication that Jim was an outstanding student at either Groton or Princeton. He seemed to be a self-achiever.

Meanwhile, his two brothers, Sydney and Robert, were starting their careers as inventors and designers starting with classy automobiles. During his first year at prep school, Sydney built a gasoline engine and designed a car that he drove when he attended Harvard. In 1909 he designed and built the first hydroplane motorboat. After several years designing special airplanes for the Navy, Sydney devoted much of the rest of his life as an innovative yacht designer.

Robert started work in France working for Peugeot and Renault. In 1911 he designed a four-cylinder sports car and in 1914 he won the Tour de France automobile race. After the first world war, he built the famous Breese-Paris roadsters in Paris. A few still are owned by collectors. He had patents for a variety of conveniences ranging from the first woman's razor to can openers with the magnet to hold the lid.

Jim's sister Frances seems to have been her father's favorite, and when his investments were doing well, she had private tutoring at home with Swiss governesses. Later, she attended schools in Europe where her first language was French. After living a few years in establishment Long Island, she lived the rest of her life in a bohemian lifestyle on Long Island, New York City, Haiti, and Mexico. She had a business designing contemporary rugs for architectural clients. Frances also did some modern painting and she wrote for *New York Magazine* and other periodicals.

James L. Breese, Sr., in a road race

On April 15, 1915, Jim married Marjorie Howard Gorges in San Francisco. He was 30 and she 31. Marjorie was an attractive actress born in Quebec to an English father and an Irish mother. When Jim met her while he was at Princeton, she was a department store fashion model. She was also Catholic and for this reason, Jim's father disinherited him. His brothers and sister shared part of their inheritances with him to make up for his loss.

In 1918 Jim's father turned his attention to the new field of flying by starting the Breese Aircraft Corporation. This short-lived venture made 300 non-flying training aircraft using short wings and real but low-power engines. These "Penguin" aircraft give Army trainees the feel of flying an airplane at near flying-speeds. Today, pilots train in totally enclosed electronic simulators. These simulated

James L. Breese, Jr.   Marjorie Howard (Gorges) Breese

cockpits completely replicate the sight, sound, touch, and feel of flying a real airplane.

So it was natural that Jim and his friends became interested in aviation. Through a chance meeting with Amar Johnson, who ran the Naval Reserve Force, Johnson suggested they enlist for naval aviation training. The United States had declared war on Germany and Jim and his friends could become officers rather than deckhand recruits.

Jim's daughter Ann Breese White wrote in a memoir, "With no formal direction from the Navy, Jim and his friends set up their own training program at Bayshore, Long Island, using borrowed planes. It was all great fun, but they had the strange feeling that they were somehow not quite with it. There was no contact with the Navy or with Washington, and they weren't even getting paid! Just at the point they were again considering sending one of their group to Washington, they heard they were actually going to be inspected by a Commander Albert Read, as the Navy wanted to open an aviation training center on Long Island. As it turned out, the inspection was a disaster, with everything going wrong that could go wrong, and they were convinced by Read's austere, unsmiling manner that he was displeased."

Weeks passed and nothing happened. Then they received a telegram: "Brooklyn Navy Personnel will arrive Bayshore in 48 hours to assume command, commission crew, and begin designated flight training program. Lt. James L. Breese will report Navy Dept., Washington by first available transportation to join the Liberty engine inspection group under Commander Richard E. Byrd, by order of Assistant Secretary of the Navy. Signed, Commander Read." The Assistant Secretary of the Navy was Franklin D. Roosevelt, Jim's classmate at Groton.

Little could young Jim Breese know that this fortuitous chain of events would lead to his great adventure of a lifetime.

2

Preparations for the Transatlantic Flight

The Breese family's interest in airplanes was timely. Interest exploded in exploring the possibilities of this new technology. Lord Northcliff, owner of newspapers in London, saw a once-in-a-lifetime promotional opportunity. He announced in April 1913 that his *Daily Mail* was offering a prize of 10,000 pounds (about 1.5 million dollars in 2020) to the first person to fly across the Atlantic. The flight could take no longer than 72 hours.

Only a seaplane could make that flight, and Glenn Curtiss on Long Island designed and manufactured the best seaplanes. With the backing of Rodman Wanamaker, heir to the department store fortune, Curtiss produced a flying boat, *America,* designed to fly across the Atlantic. The plane performed well after Curtiss added a third engine. In the subsequent NC series flying boat design, Curtiss added a fourth engine. This additional engine was a backup in case one engine failed.

The *America* was to take off from Newfoundland on August 14, 1914, refuel at the Azores, and then fly on to England. But on August 3rd Germany declared war on

France and the next day Great Britain declared war on Germany.

World War I provided an unforeseen boost to the transatlantic project. The British needed patrol aircraft especially to sink the growing menace of German submarines. The American seaplanes rigged for bombs and depth charges seemed the best available solution. But many of the ships carrying the seaplanes from the United States to England were sunk by submarines. It would be much better if the American seaplanes could fly themselves across the Atlantic.

In 1917, Glenn Curtiss began the design of a new plane that would be larger, travel farther, and would carry armaments. Josephus Daniels, the Secretary of the Navy, signed the papers formalizing the Navy-Curtiss partnership. Thereafter, planes from this development effort would use the designations with the first two letters NC. The initial order in 1918 was for four planes, the NC-1, NC-2, NC-3, and NC-4. They were assembled in Garden City, Long Island, and then moved to the Naval Air Station at Rockaway Beach.

These were very large planes for the time:

Wingspan: 126 feet (Four feet shorter than a Boeing 707)
Length: 68 ½ feet
Weight: Empty: 15,874 lbs., Full: 28,000 lbs. (14 tons), Useful load: 12,126 lbs.
Total fuel capacity: 11,346 lbs. (1,891 gallons)

Destiny Strikes Twice

Engines: Four Liberty V-12, 400 horsepower each for a total of 1,600 horsepower
Fuselage: Manufactured by Herreshoff Manufacturing Co. (Yacht designers)
Maximum speed, full fuel load: 74 knots (85 miles per hour)
Maximum range: Approximately 1,200 nautical miles (1,380 statute miles)

The first flight of the NC-1 was on October 4, 1918. After several months of tests, the NC-1 was flown to Washington to show her off to senior officials. One half-hour into the flight, a radiator sprung a leak and the NC-1

The NC-4. The commander-navigator is in the nose, the pilots are in front of the wing, and the engineers and radio operator are behind the wing.

had to be brought down on the ocean. Waves were running as high as 10 feet and it was a real test of rough water capability. After it was repaired and shown in Washington, Admiral Montgomery M. Taylor approved the design. On October 31, Commander John H. Towers, one of the first Navy pilots, proposed a transatlantic crossing using the NC planes. Hardly a month later, the Armistice was signed, ending World War I.

In 1919, seeing the project sinking, Towers pressed ahead, moving his proposal up the chain of command to Franklin D. Roosevelt, the Assistant Secretary of the Navy. Roosevelt was receptive and enthusiastic. When Towers told him there were parts for three more NCs, he said, "Why not get them all to Glenn Curtiss at Garden City and let him assemble them? Then we can make the flight to Europe just for the cost of gas." When the NC-2 was completed, Commander H.C. Richardson took Roosevelt on a short but bumpy flight. When he was back on land, Roosevelt was airsick but full of enthusiasm.

A committee was appointed to study the proposition. They recommended that a flight be made by way of the Azores, that all four NC seaplanes be flown to Newfoundland as the take-off point, and that the three best performers continue to Lisbon and Portsmouth. The plane left behind could be a source of replacement parts for the other three.

Meanwhile, Lord Northcliff renewed his 1.5 million dollar offer for the first successful transatlantic flight. Many teams were announced but serious contenders dwindled to six planes. The U.S. Navy made it clear that it could not accept prize money.

Jim Breese's daughter Ann Breese White wrote: "The British did not consider the U.S. Navy project truly sporting. One of the contending pilots, Harry Hawker, scoffingly said it was a Cook's tour across the ocean and offered to eat the seaplane that actually made it. The British also objected to the fact that the Navy flight was carefully planned, with sixty ships strung across the Atlantic beneath the path the NCs would fly to the Azores. This kind of preparation some believed eliminated all adventure from the flight. And through all the controversy, the Navy insisted it was not interested in the competition."

She continues: "Arrangements on the NC boats seem primitive to us today. There was absolute disregard for even basic crew comfort. It was discovered too late, for example, that there was no provision made for a head (toilet), an unbelievable oversight in my opinion (uninflated weather balloons were used as toilet substitutes). The plane's commander, who also served as navigator, was stationed in the bow, where he had a small table, his instruments, charts, and logbooks, and a box for his seat. Commander Richard E. Byrd had designed especially for the flight a bubble sextant and a wind drift indicator. To handle all this paraphernalia and, in fact, to see anything outside the plane required the navigator to rise through the circular pit at the extreme front end of the plane (originally a machine gunner's position). Then, standing in the wind, he would do his calculations from the sun and the stars, if they were visible. The hull was divided into segments, each compartment having bulkheads with outboard access doors through which a man could crawl fore or aft. The pilots sat on a bench facing the wind; all that separated them from the elements was an automobile-type windshield. Nine gas tanks were located amidships. Toward the stern was the cockpit for the engineers, and below that were the crowded quarters for the radio equipment and its operator. Members of the crew communicated through a metal speaking tube, or they could use headsets. The radio operator could talk to planes or ships within a radius of only about 20 miles. The crew's flight suits were lined with fleece, beneath which each man wore a regulation uniform and heavy underwear. There were no parachutes, but the crew had lineman's belts, a security measure to prevent their being swept away while climbing about the engines during flight."

Demonstration of the size of the NC-4

Jim Breese was known in the Navy as a power plant expert, especially on the Liberty engine which was used in the NC-4. Also, Franklin Roosevelt was a school classmate and boyhood friend. Thus he was a good choice for the NC-4's crew. They were:

> Lieutenant Commander Albert C. Read, commanding officer and navigator
> Lieutenant Elmer Stone, U.S. Coast Guard, pilot
> Lieutenant Walter Hinton, copilot
> Lieutenant James L. Breese, engineer
> Ensign Herbert C. Rodd, radio officer
> Chief Machinist's Mate Edward H. Howard, engineer

Jim was also referred to as pilot-engineer because he was a qualified pilot, and he piloted the NC-4 during parts of its history-making flight. Jim and Commander Read had quite different personalities. Read was a reticent and curt New Englander, a Navy man to the core (He eventually became a rear admiral). Breese was flippant and joking and he did not amuse Read. Nevertheless, they developed high mutual respect during the long transatlantic flight.

Jim was an engineer to the core. He wrote: "Many months before the start of the transatlantic flight I was sitting at my desk in one of the huge hangers at Rockaway compiling the results of several test trips made with the first of the NC boats...Out of the mass of figures finally came the answer: 1.3 gallons per nautical mile. To you this means nothing, to me it meant everything—it meant we could cross the Atlantic. The thrill and sensations were only comparable to those I had when we actually completed the flight. For my part, I had crossed the Atlantic that night, and to be a member of the crew that accomplished the feat and to have the added good fortune of being the first to cross was of course not in my original calculations."

Because of several accidents, the NC-4 was regarded as a jinxed aircraft. Just a few hours before the scheduled take-off, the chief mechanic, E.H. Howard, stepped too close to the rotating propellers and lost his hand. On the way to the first aid station, he called back to the horror-stricken crew, "I'll be right back. Don't go without me." Nevertheless, he was replaced by Eugene S. Rhodes. This decision was devastating for Howard who had worked on the NC engine installations from the beginning.

Finally, on May 9, 1919, the NC-1, NC-3, and the NC-4 were ready for take-off to start their historic flight. The NC-2 was left behind to continue as a source of replacement parts. It was a clear day and they planned to land for the night in Halifax, Nova Scotia. After they took off, Franklin Roosevelt radioed a message, "Delighted with the successful start. Good luck all the way." In the next chapter, I will let Lieutenant James L. Breese tell much of the story about their adventures culminating in the NC-4 landing in England.

Destiny Strikes Twice

U.S. Navy

U.S. Navy promotional painting for the NC-4 flight by Ted Wilbur. Jim Breese is the second from the left.

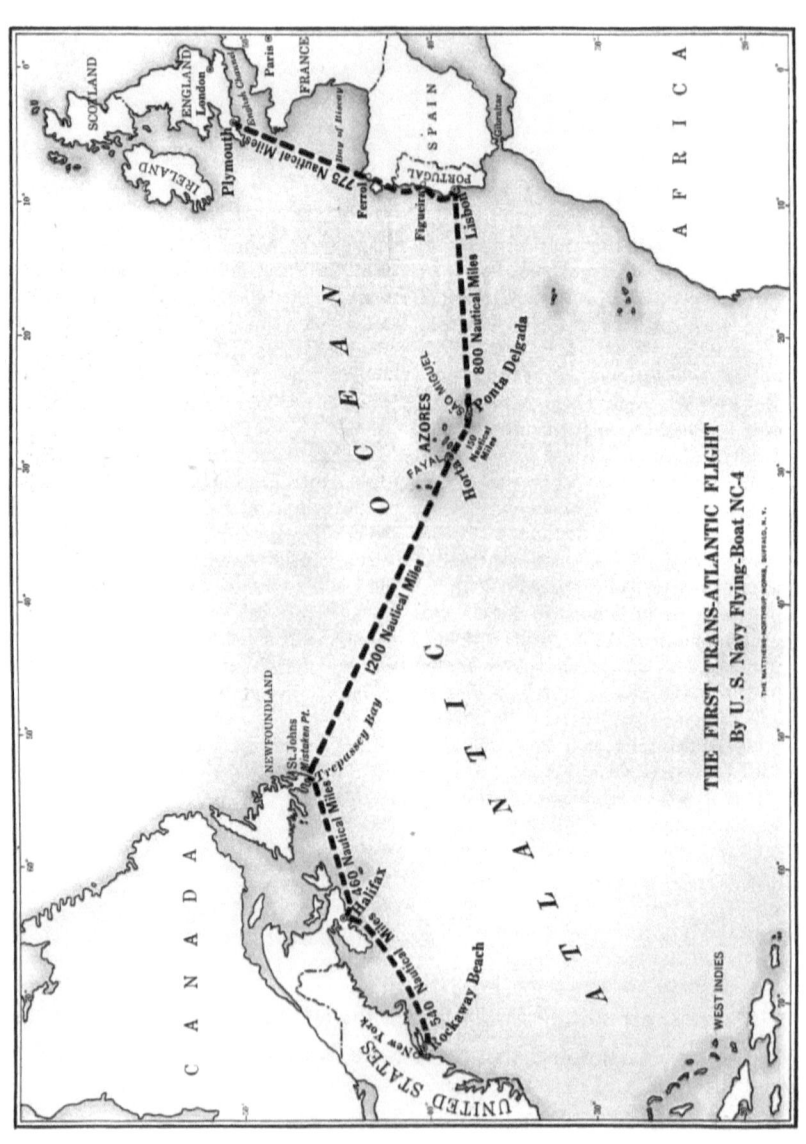

The route of the NC-4

Destiny Strikes Twice

3

The First Flight Across the Atlantic

There's no way I can tell the story better than Jim Breese. He was right there, impressionable, and full of excitement. Below is his complete flight diary. His engineering report and other technical data are found in *First Across*. I should add that Jim wrote to his wife Marjorie when he reached the flight's first stop in Chatham, Cape Cod, Massachusetts: "...I hope you didn't worry about me because you know I always bob up again even tho everything doesn't go the right way. Goodbye and lots of love..." This is the mind of an achiever.

<u>Jim's diary</u>
<u>Friday, May 9, 1919</u>
<u>On board the NC-4</u>
<u>Rockaway, Long Island, Naval Air Station</u>

At ten o'clock this morning the three NC machines were standing, waiting to start on the first lap of our journey. NC-1 lay fifty yards up the beach, Lieutenant Commander Bellinger standing in the navigator's pit, anxious and eager for the start; NC-4 in the center with Lieutenant Commander Read, and NC-3, the Flag Ship of the Squadron a little further down the beach with Commander Towers. NC-2 has not made ready, for at various times parts of this machine were removed and used to replace damaged parts on 1 and 3. With their noses on the beach, their motors revolving in lazy fashion, they showed no signs of nervousness on this great venture, but seemed like fine animals ready and eager to push off, and, as at a racecourse, the crowd came near and little groups formed around the bow of each ship, shaking hands and giving the last words of farewell. Each member of the separate crews was presented with a four-leaved clover to bring him good luck.

At 10:03 the order was called, and the men in high boots, wading out into the water, turned each machine around until the bow pointed out to the sea. No. 3, the flagship, opened her throttle, taxied through the water to a suitable position for taking off, followed by no. 4 and then by no. 1. Almost immediately full power was thrown on and no. 3 rose from the water in a cloud of spray, closely followed by no. 4 and no. 1, and in close formation we started on our course along the shores of Long Island. As we proceeded, we spread out and the distance between us grew greater.

The engines are working beautifully, the air is smooth and the world seems very rosy. The land seems to sink from beneath us and with it all the petty annoyances, delay and troubles of preparation and starting to disappear in the distance.

We are up—we are off! and here in my little cabin, which I share with the radio operator and the mechanic, I can sit down and write the notes for what will be, I hope, the story of another success for our Navy.

Since leaving Montauk, I have been giving the machine a thorough inspection. I found a lot of water in the gasoline tanks, and went forward on my hands and knees along the starboard passage, about two feet by twenty, leading from the cabin to the pilot's seat, drawing off the water as I went. On reaching the pilot's compartment, I relieved one of them, and taking the wheel took my course from the navigator and flew for an hour. Then we had a sandwich and a drink of coffee, and I took advantage of my place in the forward compartment to have a smoke. For this, of course, is forbidden aft the gasoline tanks. We are now passing our Cape Cod, headed for Halifax, and are having our last sight of the U.S.A. We picked up destroyer no. 1, but we see only one of the NCs. The other, I think, is to the south of us, as I last saw him taking a slightly different course from ours.

Saturday, May 10
Chatham Naval Air Station

Just as I finished writing the first entry, there was a hurried call from the pilot. I went forward to see what was the trouble and found that that the oil pressure in the rear center motor was dropping. This would indicate one of two things: either we were running out of oil, or the oil line was becoming clogged. In either case, the motor would be ruined if we continued running it. After exchanging a few hurried notes with the pilots, I advised them to shut off that engine and continue on three engines. This they did, and we sent out a wireless, telling the others that we were OK relying on three engines. No sooner had we sent this message than the forward center oil pressure dropped, the connecting rod broke and came out through the side, breaking the water pipe and we were forced to come into a glide for a landing on the sea about one hundred miles off Cape Cod.

A long swell was running and while we were making an inspection, most of us were pretty seasick. I found the oil nearly gone in the center tank and the feed in the rear engine stopped up. Commander Read decided no to spend any more time trying to get it fixed and we started taxiing with two engines, headed for the Naval Air Station at Chatham, where we knew we could procure a new Liberty motor. An all-night run brought us there at daybreak. A wonderful piece of navigation: Imagine landing somewhere in the open ocean, setting your course at an unknown speed, running all night, and at day-break finding yourself just where you wanted to be.

Tired and hungry, we landed at the Station and soon a squad of men was taking out the old engine. I went off to the officers' quarters and slept until 2:00 p.m. and then took charge for the night. We worked all night and by noon the next day were ready to set off and catch up with the others. But no! A message came from Washington: "Weather unfavorable for flight to Halifax. Storm approaching." We secured the plane and now we are sitting around, day after day, getting hourly weather reports and reading the progress of the other ships as they passed Halifax and arrived at Trepassey, and worrying ourselves sick for fear they will get away before we are ready.

Wednesday, May 14

Off at last and headed for Halifax. The engines worked rottenly all morning and the new center forward engine is vibrating so badly that it is dangerous. But we are off and are going to keep going. It is just noon and the station and shore outline are very pretty in the bright sun, with the blue water and white sand. We once more passed over Destroyer no. 1 and except for the terrible vibration, I think we'll make Halifax. I had the pilots throttle the forward center engine to twelve hundred revolutions per minute and the other engines are working about fifteen hundred. We are traveling about seventy-five knots. Ensign Rodd, the radioman, is talking to everyone around here, as he says he is working Bar Harbor, Halifax and Trepassey Bay besides talking to the destroyers below. I have just had a lunch and a turn at the wheel and a smoke up forward with Commander Read. There is a leak in the gasoline line which is annoying me but I will let it go at present. We have picked up all the destroyers, and are now following the coast of Nova Scotia. Except for the lighthouses, and little fishing villages, it is not much to look at. I have just read the morning paper, which I brought from Chatham, shaved and washed with warm water drawn from the radiator overflow system and feel altogether quite fit. We will be making Halifax soon, so my next entry will probably be written on board the *Baltimore*, the mothership of that station.

Night of May 14
On board the U.S.S. *Baltimore*

A hearty welcome and a hot supper greeted us after we had landed and tied up to the stern of the big battleship. After calculating our gas and oil consumption, giving orders for refueling and some minor adjustments, we all turned in. Our gas and oil figures are exceedingly high and not at all what I figured out. I attribute it to the poor forward center engine and dirt in the line to the starboard engine.

Thursday, May 15
On board NC-4

"Off again on again gone again Finnegan" was the procedure this morning. We broke a starter warming up at eight o'clock and took two hours to replace it. When we finally got off, the oil pressure was bad and the starboard and center engines were not hitting right so we landed in a delightful little bay about fifteen miles from Halifax, and spent two hours cleaning out the lines. But we are "gone again" now; the engines, except the center forward, are hitting well, and with any sort of luck we will make Trepassey tonight.

There has been nothing much to write about all afternoon, just the usual routine, taking the readings of the instruments every fifteen minutes, watching the engines for leaks, eating occasionally and having a smoke and chat with the pilots and navigator now and then. I say "chat" for lack of a better word. It is, of course, impossible to talk because of the noise of the engines so these communications are carried on almost entirely by means of signs and gestures, and one who has never been forced to try it has no idea of the details one can convey by this means. In fact, we were often able to relieve the monotony by swapping, or so to speak "gesturing" a funny story or two. It is getting colder and the shoreline is becoming very bleak and rugged. Snow patches on the northern slopes are noticeable and there are very few signs of habitation.

But here we have something to worry about. Ensign Rodd, our radio operator, who sits always at his desk recording instantly any messages he can catch from the air, became much excited and began writing rapidly in his logbook. He had caught the message to the destroyers along the line between Newfoundland and the Azores. They were receiving instructions to be in position tonight. That means the others will start without us. Doggone the luck! Another rush message has just come in. "C-5" (the Navy dirigible) broke loose last night from its moorings north of Halifax. No one on board no casualties, drifting in a northeasterly direction. Some time ago we closed the hatch to keep warm, but curiosity or something made me open it to look out; there, about six miles off, I saw the C-5 drifting along about two hundred feet off the water. There is nothing we can do, so we are proceeding on our course. It is getting terribly cold, there are icebergs below. The water from a leak in one of the pipes is forming icicles on a brace wire. Trepassey must be very near. The signals from the *Aroostock*—the mother ship there—are very loud and we have just received a message to be on the lookout for NC-1 and 3, as they are preparing to leave. For the love of Mike! but perhaps they will let us start anyway tomorrow after we have changed our forward engine.

Thursday, May 15
On board the *Aroostock*

Never will I forget our landing at Trepassey Bay. The mother ship lay at anchor, and the clear blue water was breaking in white caps as we glided down, just skimming over the tips of some houses. We took to the water and just then I saw NC-1 and 3 coming back again up the bay. What had happened? They hadn't gone after all! Our spirits rose as we taxied a mile or more to our mooring, the foam flying with 1 and 3 roaring along beside us and making us feel quite welcome. We moored the 4, a boat

Destiny Strikes Twice

came out for us and once more in the friendly warmth of the ward-room, we met our companions whom we had left at Rockaway.

Lieutenant McCullough, pilot of no. 3 kidded me, calling me a punk engineer, as I was the only one who had engine trouble, but I was so glad to get there, that nothing mattered. He could have called me anything he liked.

The engine we had put in at Chatham was not a new one, and was put in only as a temporary matter, to get us to Halifax. A gang is now at work, changing engines once more and refueling, so by noon tomorrow we should be ready to start with the other ships.
They have tried to get off with too heavy a load of fuel, so they will reduce and we will all go off together. I have just had a session with Lieutenant Commander Richardson, who kept the data for 1 and 3, and our information compares pretty well. We are going to load up with 1600 gallons of gas. This may get us to Ponta Delgado in the Azores, but if we are short we can stop at Horta, two hundred miles nearer.

On board, the *Aroostock* everybody seems excited. in the wardroom little groups of specialists, engineers, navigators, radiomen, pilots are discussing their own special problem. Everybody has information and weather reports, press dispatches, telegrams, etc., which seem to be scattered all over the place. I am going to bed and get a good night's sleep.

Lt. Cdr. A. C. Read (left) with Lt. Cdr. Robert E. Byrd (not on the flight) and Lt. Walter Hinton in the pilot-navigator's open-air cockpit of an H-16 seaplane (very similar to the NC-4) to refine methods and equipment for the transatlantic flight.

<u>Night of May 16th</u>
<u>On board NC-4</u>

The strain of getting ready in time to get off is telling on me, and if it weren't for the exhilaration and the cold I should be all in. We worked frantically to get the new engine in and ready for this afternoon. At four o'clock the other crews started for their boats and still we were not ready. Messages began to come in from the flagship commander Towers, "How soon will you be ready?" We answered, "In half an hour." So the others cast loose and began taxiing around us. Our new engine was full of oil and wouldn't start. It was a terrible half hour to see the others all ready and our engine not yet started. I tried every wile I knew to coax a balky gas engine into starting, but a couple of sputters was all I got. Finally, as the others were pulling out into starting position, I thought of one last trump card, and went below and raised the ignition circuit from its normal eight volts to twelve. I was taking a chance of burning something out but the moment was critical. I came back on deck, the pilots pulled the starting button and blinkety, bang, bang, off she went. Just an old-timer. The other engines started easily and we shot out into the bay. After a few minutes' warming up, we saw the others start, so the pilots gave her the gun and in two minutes and fifteen seconds we were off the water. NC-3 failed to get off so we circled the bay and landed. I think they took off one man. We all then started again — again we got off first, 3 second, and 1 last. I have seen very little of them during the last two hours.

The big hop is on, the sun has set in a ball of red behind us, and the moon is rising off our port bow, the engines are running beautifully, we are snugly shut in our little cabins with lights on, as we already have passed over two destroyers, everything looks rosy. No. 1 is about on a line but behind us and no. 3 seems to be to the south and a good way below us. Our cabin is brilliantly lighted and rather snug. With the hatch closed we get none of the wind and can be very comfortable. Within the space of eight feet by ten feet, we have in one corner the deck for the radio operator, on which his instruments are mounted. Here he sits hour after hour, taking down messages he can catch. The mechanic is seated in another corner with his feet up, his chair tilted up, sleeping peacefully when not on duty. As for myself, I sit on the toolbox between the radio operator and the instrument board, so that I can conveniently keep a record of the instruments and also glance at the messages, in and out, as they come to the deck. The two pilots' seats are close together, and are far enough up to allow the pilots to see out over the deck. In front and below there is a small cabin where the pilots can sleep and store their clothes—a space about one half as large as the engineers' cabin. Directly ahead of this compartment, and connected with it, is the navigator's chart room. This is just large enough to hold one full-sized board and a small rack for the book rack for the books of reference. Of course, neither the navigator nor the radio operator has any opportunity to sleep. In the case of the pilots and engineers, where there is a chance for relief, there is ample room to sleep comfortably in either the pilot's cockpit or in the passageway connecting the fore and aft cabins. As a matter of fact, except for the mechanic, few of us slept much, for the work was too interesting, but at least half of us could have slept at any time all the way over. We took with us some sandwiches and a thermos bottle filled with coffee, besides the regular emergency ration for a crew of six, but not more than a quarter of this food was eaten, and the emergency ration of NC-4 was not touched. We were provisioned for twenty-four hours. When it comes to a question of washing and shaving, it so happened that we had installed an auxiliary water supply system for the radiators; this made it possible to draw off hot water in our cabin, whenever we wanted it, so washing and shaving became a matter of course. Even Commander Read, who was exposed, did not suffer from the cold in the flight to the Azores; we were all well-supplied with warm flying suits.

I spend my time reading the radio messages, keeping a record of the oil, water, and gasoline gauges, and going forward now and then to take a turn in the pilot's seat, while he goes below and has a smoke and perhaps a little nap.

I have just awakened from a fine two hours sleep. I feel much refreshed and ready for anything. But nothing seems to happen tonight. The engines are running like clockwork—tongues of flame are shooting out on the side of each engine—no leaks and no vibration; a beautiful moonlight night and half way to the Azores!

What more could a man want!

Radio messages indicate that we are ahead of the other two, which are coming along regularly, but about a half-hour behind us.

The radio apparatus on NC-3 does not seem to be working as well as it might. Daylight is coming and with it a cloudy sky. It doesn't look bad, everything is going fine. We are averaging 80 to 85 gallons of gas an hour. At this rate we can make Ponta Delgada.

Early Morning, May 17th

The fog closed in on us and we had to go up to three thousand feet. It is terribly pretty up here, but very hard for the navigator to get his drift or to sight the destroyers. The clouds are getting worse. At one time we spun completely around but now we are doing better.

We have found a thin layer between fog banks and are flying in on that. The NC-1 is coming along regularly, but we have not heard from the NC-3 since destroyer 17. The fog is much worse as we are flying now close to the water. If we are on our course we are liable to run into land pretty soon, and if we are off it, I don't see how we will get back. I wish this fog would lift. Three cheers! — land sighted off our starboard bow. I don't know what it is but it looks good to me.

Monday, May 19th
On board U.S.S. *Columbia*

We came down shortly after we sighted land and taxied up into the shelter of a bay. There seemed to be nothing there but one lonely church. Commander Read guessed where we were and told the pilots to take off again. We flew around the next point, and there, a half-mile ahead of us was the *Columbia* and the little town of Horta, island of Faial, in the Azores.

We taxied in, boats came out to meet us, and as we slid into the harbor, I hoisted the American flag on NC-4 as a commissioned ship of the U.S. Navy entering a foreign port.

We got the usual cordial reception on board the *Columbia* plus something more. It made me feel that we had accomplished something of which the Navy was proud and that it was only the forerunner of future trips. I believe Columbus put off from the Azores to come to America. It seems quite fitting that we should repay the compliment by this return voyage of discovery. (note: Columbus stopped at the Azores only on the return from his 1492 voyage to America.)

We ate three or four steaks and then sleep overtook all the crew except myself. I did not feel sleepy, so I attended to refueling and moored the machine for the night. An entertainment was arranged for us at the local theater and we managed to go through in pretty good style, meeting all the prominent citizens. We are now standing by waiting for clear weather so that we can proceed to Punta Delgada and thence to Lisbon. No. I's crew came aboard the day after we landed. They were lost in the fog and came down in open ocean to get a bearing. In making their landing in twenty-foot waves, they broke a wing tip float and finally a lower wing panel gave way. They were picked up by a passing tramp steamer at night, after being tossed around all day in a forty-knot gale. The tramp took the plane in tow and brought the crew on to the *Columbia*. During the night the tow line gave away, but the plane was recovered and taken in tow by a destroyer. It was later lost for both wings were torn off by the high seas and finally it turned over and sank.

All this time we had no word of no. 3. Destroyers were searching for her around the position she was last heard from, but no news at all. Finally a message from Ponta Delgada said, "No. 3 sighted five miles off the harbor proceeding on the water under her own power." So she was safe after living at sea seventy hours. I guess the crew must all be in; we will get the whole story when we arrive tomorrow if this weather will only let up.

<u>Tuesday, May 20</u>
<u>On board NC-4</u>

We have just left Horta. The volcano on one port stands up so high that it makes one feel very small; clouds surround the top, but snow is visible on the side. The air is so "bumpy" that it is hard to write. The engines are humming along without a miss and the clear blue sky looks fine after three days of fog. These last few days have been full of alternate thrills and disappointments. Our arrival was unique and exciting. There came twenty-four hours of worry for no. 1. Finally her crew was brought in safe. Then came two days of worry over no. 3. We had no sooner heard that 3 was safe coming into Ponta Delgada under her own power, than there came a report that Hawker, in the Sopwith, had landed in Ireland—thus taking all our chances of making the first trans-Atlantic flight.

The relaxed and happy NC-4 crew after the long flight to the Azores. Jim Breese is the second from the right.

Larry Kilham

Friday, May 23
At Admiralty Office

Here we are, still held up in the picturesque town of Ponta Delgada. Our arrival Tuesday was a gala affair. The whole town turned out to meet us. We came ashore and a battery of "movie" men and kodaks took us from every angle. It was all very exciting meeting no. 3 crew. Hearing their tale of having been shipwrecked for seventy-two hours. How they, too, came down in the fog to get their position — broke up on the landing and could not get off again. Finally after being battered about for two days, they got enough clear weather to find their position and by controlling their drift, they finally got to this port, their last port, with only one motor running.

The afternoon of our arrival, the governor had a reception for us, where we met the prominent citizens of the town and the high military officials. That night a dance at the Admiralty closed our official reception.

We were to start the next morning at daybreak. The machine had been looked over and filled up with gas and oil and all was set. At five o'clock we were on board — the engines were started and we taxied our way out of the harbor. There were some swells running but not a breath of wind, and after making two unsuccessful attempts to get off, we returned to remove some gasoline. This was a slow process, and when it was finally completed we made another trial, but this time a small piece of dirt clogged up

Destiny Strikes Twice

one of the jets in the carburetor. By this time, it was too late to start, so we moored for the day.

Since the weather conditions have been steadily worse. This is the type of report we got: "Weather report for 4:00 p.m. 5/23/19."

> Barometer 30.12 inches, unsteady
> Wind S.S.W. 25 miles
> Weather clearing but unsettled, sea moderate.

Forecast: Present report indicates that Saturday will be unfavorable for flight to Lisbon owing to direction and strength of wind at various points on course. Sunday now seems the probable day. And so it goes each day as at Chatham and Horta, we seem to be held up again.

<u>May 27</u>
<u>On board NC-4</u>
<u>Bound from Azores to Lisbon</u>

Well, here we are, slipping through the air again at our usual rate, with a good stiff wind from behind. At Ponta Delgada it rained and blew steadily for six days, just as it did at Chatham. We tried for an early start this morning, but luck was against us. At five o'clock we had our motors going and we were jockeying around for a start when we noticed the first engine was not acting right. The carburetor had been removed during the night in order to free the throttles which were crowded and sticking after a week of rain and storm. It took us until nine o'clock to get fixed up. We made a splendid get-a-way against a strong wind and heavy sea.

It is a beautiful day, but there is nothing to see or do; except for the destroyers which pass under us every forty-five minutes, there seems to be nothing in the world. We have our own little universe, which seems to have its limits just where the sound of our engine stops.

It is evening now, and the sun is setting below our "tail" just as it did at Trepassey on the night of the "big hop." We have checked off all the destroyers except for no. 14 – the last one. When we pass him there is only Lisbon to come and if we land safely (it will probably be dark), we will be the first to have flown across the Atlantic.

We have sent word ahead by wireless, that we will arrive about 8:30, our time; about 6:45 their time, and they are to have searchlights out and will do all they can to help us. It is certainly a novel experience to be landing in a foreign port at night with no knowledge of the place at all.

<u>May 29</u>
<u>On board U.S.S. *Melville*</u>

Our entrance into Lisbon was most picturesque. As soon as we sighted land a thrill went through us, and as I looked forward over the deck, I caught Commander Read's eye, for he had turned around to give some

Destiny Strikes Twice

directional instructions to the pilots, and his expression so exactly expressed my feelings, that I am sure between us we must have registered satisfaction 100%. How different it was for Columbus and his crew when they sighted land! They danced on deck and yelled for joy, but we poor dwellers of the twentieth or "movie" century could only register facial expressions and fly on. And so we did and as we came nearer the land took shape, and soon there was a shoreline and then a beach, and then the mouth of the river; then neat, square lined white houses with winding roads and then forts and castles and boats and then a city and a wonderful harbor. Flying at 1500 feet, we cut the motors over the city and made a wide spiral glide over the bay and landed within a quarter-mile of the *Melville*—the mother ship. In the usual way, we taxied up to our buoy, and when we finally shut off our engines and were about to take notice of the thousands of silent curious little sailing craft which were crowding about us, we had the surprise of our lives, for out of silence the most stupendous noise came to our ears. The Admiral's barge took us to the flagship *Rochester*, and we were hurried up to the quarter deck under a battery of searchlights, for it was now nearly dark, and there—still in our greasy flying clothes—were received by the Portuguese high officials of both the State and the Navy. No praise was too great for us, apparently, and there, on the deck of the battleship, in the glare of the searchlights with the native band playing American tunes, we were decorated with the highest order the government could bestow—the Cavaliers of the Tower and Sword. Gayly dressed women and well-uniformed men made up the enthusiastic assembly on board and the weird bluish light from the searchlights made all appear like a glorious dream. Then handshakings began, and we soon realized it was no dream, but that we were really across the Atlantic. The following day the NC-4 was gone over for troubles, but none were to be found, she was refilled and made ready for a start when the time was propitious.

U.S. Navy

Taxiing in Lisbon Harbor

We spent one more day in Lisbon and received on that day twenty-four hours of solid entertainment. On the morning of the third day — worn out by so much attention we finally got away before sun up and looked forward to a long day of rest on this last leg to Plymouth.

Destiny Strikes Twice

We were gassed up for only twelve hours' flight. We were heavily loaded, and got away in good shape. There was little wind blowing, but a good tide in our favor. After rising sufficiently we swung out over the bay, and then flew directly over Lisbon, where we had been treated so hospitably during our stay, and whose officials had particularly requested that we do this, if possible, when leaving. The coast of Portugal is not particularly interesting to look at as it is rather rugged and barren, but it is so infinitely more interesting when the endless miles or just plain ocean we have been passing over, that I spent most of my time looking out of the hatch at the shore some five miles away. There were many low-lying clouds, and I noticed for the first time the phenomena of the complete circle rainbow. It is quite the prettiest rainbow I have ever seen and occurs when the machine lies between a cloud and the direct rays of the sun, provided these rays are not more or less horizontal. This normally would cast a shadow on the cloud and the rainbow makes a complete circle around it.

We had not been gone a great while from Lisbon when we noticed a leak somewhere in the port engine. I diagnosed this as a gasoline leak, so it was decided to land and repair the trouble. In order to land in smooth water, Commander Read headed the ship to the nearest inland waters which happened to be Mondago River, and here we glided down and took to the water. After the engines had been stopped and an inspection made, it was found that the leak was not gasoline but a slight water leak around a cylinder jacket.

This was easily stopped and we were ready to leave again. But no such luck! As we pushed out into the water, we ran into a sand bar. I got overboard, suitably clothed for the water. After a little persuasion, we got the good ship off the bar and then ran her to the shore where we secured her with a line. It was obvious that the tide had fallen and as we were blocked from getting out to sea by a bridge across the river, we had to wait until the tide rose again at 2 p.m. It was a beautiful day; the sun was shining, the sky was blue, the fields green, and the air mild, and on the river bank about two hundred native children had gathered to look at us. I was rather glad we came down.

Ensign Rodd immediately got busy with his emergency wireless set and got into communication with the nearest destroyer. The information was relayed out and I heard afterward that Washington was informed at 10 a.m. which was about the time in Portugal that the event occurred.

At two o'clock the tide had risen sufficiently for us to take off, and with a mighty roar and a hop, we sped down the river, cleared the bridge and once more put off down the coast.

It was obvious that we could not make England that night, and accordingly we received orders from Commander Towers at Plymouth to proceed only as far as Ferrol in Spain that night, and continue the next day. This we did — a destroyer meeting us by arrangement at Ferrol for the night.

Destiny Strikes Twice

A good start in the morning put us across the Bay of Biscay by early noon and after putting in at Brest—where we circled the harbor and paid our respects by wireless to the *George Washington*—we set out for Plymouth.

The fog was thickening and as we passed Ushant light it became so dense that we had to come down to within fifty feet of the water. For three hours we flew through the fog and then, wonder of wonders land rose up. We looked down to see if we could recognize it and there we were, directly over Plymouth Harbor! In other words Commander Read had navigated two hundred miles through fog and crosswind, and we had come out exactly at his destination! This was not luck, for we had done it twice at Chatham, then at the Azores, and now for the third time at Plymouth.

The harbor was gay with boats and ships flying flags—the shore was black with people, and in the air, British seaplanes swarmed in to welcome us like a flock of birds. We made one great circle of the harbor to get our bearings and then Lieutenant Stone—our first pilot—brought us down a half to the left and we landed almost in the midst of the small crafts that had come out to meet us. A fitting finish to a historical trip.

Our welcome in Plymouth was splendid, wholehearted and enthusiastic, the people of the city burst out to welcome us when we came ashore and the Lord Mayor received us at the very stone from which our Pilgrim Fathers set out three hundred years before.

Well, the trip is over. The Navy has accomplished its purpose — the Atlantic has been crossed by air and the NC-4 was fortunate to be at the finish.

This ends the narrative of the NC-4 but not the trans-Atlantic flight; for the narrative alone was only an incident in the flight and the whole flight was the result of the co-operation of every department of the Navy. The flight was planned and thought out in every detail by Commander Towers. Nothing was left to chance. He was told to put a seaplane across the Atlantic by the new route, the air. The Navy believed it could be done, or wanted to know the reason why not.

In the data obtained and the experience gained, the Navy now knows not only that it can be done, that they have done it, but that when they want to repeat the performance, it can be accomplished with comparative ease and certainty.

To the Navy goes the first credit for having crossed the Atlantic; to Commander Read for bringing the NC-4 through when the elements had gone against him.

Mechanically and structurally, the NC-'s were perfected before leaving Rockaway, so there is nothing particularly remarkable in the fact that no trouble was experienced.

Everything functioned — and this alone was responsible for all the machines not coming through together (note: this sentence is what he wrote but his meaning is not clear).

/g/ James L. Breese, Jr.
Lieutenant, USNRF
Engineer NC-4

The grand welcome to Plymouth, England, May 31, 1919

During the stopover in Lisbon, the NC-4 crew received the Portuguese Order of the Tower and Sword award. In London, the officers received the Air Force Cross. On May 23, 1930, President Herbert Hoover presented Stone and the other members of the NC-4 crew with gold medals, specially designed to commemorate the NC-4 flight, in the name of the United States Congress.

President Woodrow Wilson congratulated the NC-4 fliers in Paris where he was working on the World War I peace treaty. He said in part, " The entire American nation is proud of your achievement. I am glad to see you and to shake your hand, and I am glad to give you my warmest congratulations."

Destiny Strikes Twice

Charles Lindbergh, after his first solo crossing of the Atlantic in 1927 commented, "When I think about it logically, I know that I had a better chance of reaching Europe in the Spirit of St. Louis than the NC boats had of reaching the Azores. I had a more reliable type of engine, improved instruments, and a continent instead of an island for a target. It was skill, determination, and a hard-working, loyal crew that carried Read and his NC-4 to the completion of the first transatlantic flight."

Wives of the transatlantic fliers await the return of their husbands in New York. Mrs. Breese with the author's mother, Frances, age 3, are on the far left.

Awards received by James L. Breese, Jr.

Destiny Strikes Twice

4

The Chicago Interlude

If Jim Breese was living the good life in Southampton, why would he take his wife and two baby daughters to the midwestern manufacturing hub of Chicago? He later wrote to my father, "My choice of Chicago, which I hated as a place to live, was merely because it presented the best of facilities for getting my inventions into production." Jim had inventions on his mind at least since he started his invention notebook in 1909. Most of the concepts there involved ignition systems for combustion engines. Perhaps he thought back to when he started the NC-4 with his jackknife bridging to a higher voltage battery. The rest of Jim's business life and success will be based on the insights and technology he developed while he was in Chicago. His move was in the early 1920s, and he settled in the northern lakeside suburb of Lake Forest.

Jim began employment as an engineer with Nohol Company of America, a steam car company. They built one car but it was too expensive. It was a tough business to succeed in. Steam car companies started and folded all through the 1800s. By the 1920s there were still over 20 hopefuls but the power advantage steam had was overtaken by the practicalities of the gasoline engine.

Jim developed a key insight at Nohol. Their car had an oil burner with automatic controls to heat the steam boiler. Jim saw that he could apply the same principle to the design of a home space heater, water heater, or oil furnace. An oil burner with automatic control would be suitable for use in a wide variety of homes and businesses. He perfected his understanding by studying a candle flame. At a particular temperature, the flame could vaporize just enough wax to power the flame without creating smoke. This principle could be applied to the home furnace and other oil burner applications.

With the demise of Nohol, it was time for Jim to start his consulting business, Breese Engineering Corporation. This gave him cash income and important business relationships while he developed his inventions further. One of Jim's engineering and business advisors for many years that he met at this time was his patent attorney, Norman S. Parker, of the firm Parker & Carter.

Jim realized that he had a chance to develop and patent the important mechanisms of oil burners. We hardly realize it today but before the 1920s people generally heated with coal. It was dirty and cumbersome to move. Then they had to put up with the coal dust and flying ashes. Oil burners could provide clean, controllable heating.

He also foresaw that his simple but sophisticated oil burner equipment could open up oil heating to millions at low-income levels. He designed his burners to sell at low prices in mass markets. This would be appreciated in The Great Depression which started in 1928. Ironically, in later years his most important customer would be the U.S. Army who also appreciated his burners for their simplicity, ruggedness, and low-cost. The Army used them primarily for tent heaters.

The basic principles of the Breese Burner are shown in the drawings that follow. The basic element is the "pot" with holes perforated all around at about four levels. Fuel oil is fed on to the bottom of the pot from where it evaporates and travels upwards as a gas. Air is fed into the pot through the holes either by natural draft or from a blower. After the air-gas mixture is ignited, flames burn all over the air-gas interface. The heat can be lowered by lowering a flat metal ring from the top of the pot down to near the bottom which causes less air to be fed into the combustion area. The fuel flow can also be controlled in an adjacent fuel oil reservoir. Either of these control functions can be done by thermostat for automatic heating to a set temperature.

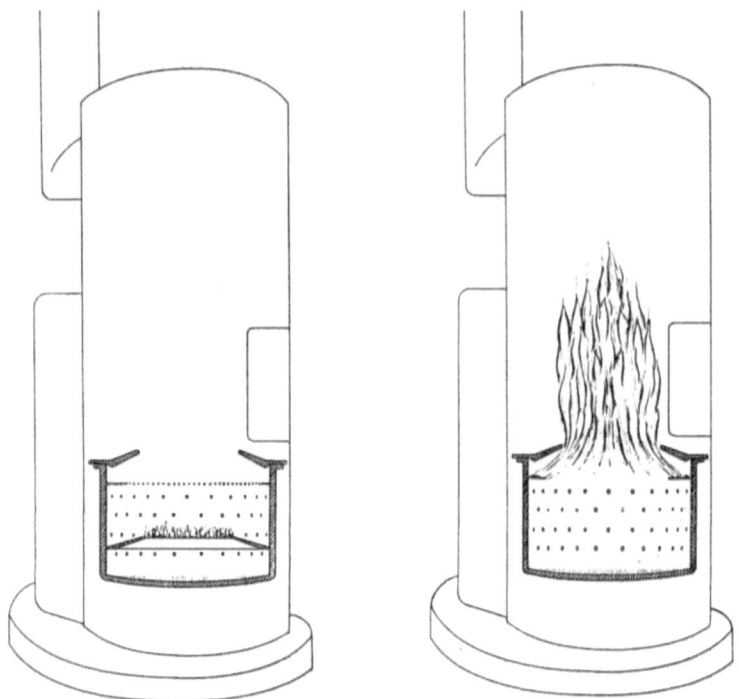

Breese burner in pilot light and full heating modes

For the next thirty years, Jim received over 130 patents for all kinds of variations and improvements to the pot-type oil burner. He listed Breese Burners "firsts" for the time he was in Chicago:

1924 – First pot-type space heater

1925 – First float control with vented stem

1926 – First pot-type furnace burner with a pilot and thermostatic control.

Destiny Strikes Twice

One of Jim's earliest patent drawings, for a Breese burner stove patent received in 1926, is shown below. This is the ornate space heater found in many homes, shops, and restaurants in the mid-twentieth century. At the left is the fuel oil reservoir with a float flow control and to the right is the combustion pot with five circular rows of air feed holes. He received 23 patents while he was in Chicago.

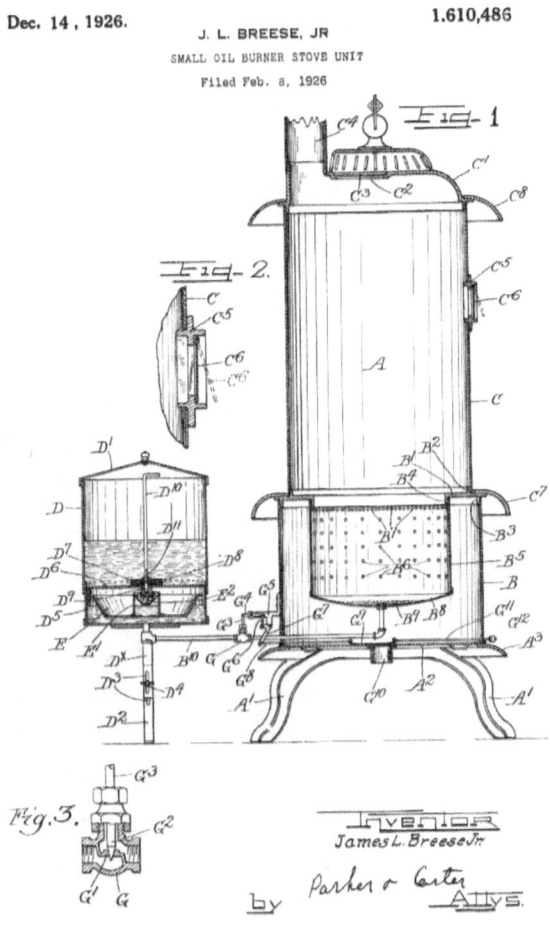

Meanwhile, Jim's family had settled into their attractive suburban home in Lake Forest. The mother of his wife Marjorie moved in. Several years earlier her husband, a retired British army officer, disappeared. Lake Forest was not a convenient commute for Jim to his office in downtown Chicago.

Marjorie set up a nursery school in space she added to the back of the house. This was convenient for the neighbors who were young families, and Marjorie enjoyed educating children. Later, after they moved to Santa Fe, New Mexico, she would set up a one-room schoolhouse.

Jim and Marjorie had two children while they were on Long Island, Frances, my mother, and Mary. Mary was always called "NC" after the NC-4 flight which occurred four months after her birth. In Lake Forest, they had their last two children, Ann and Jim. All their children in later years contributed their recollections and opinions for this book, and I have quoted at length from Ann's writings about the NC-4 adventure and life in Santa Fe.

My mother's favorite recollection was that Lake Forest, a normally quiet town and a haven for the wealthy, was also a favorite residence for the mobster chiefs. Prohibition had begun and they wanted a good place to raise their families out of public view. She recalled, however, that on more than one occasion, big black cars would come screeching around the downtown blocks, and their riders would be brandishing guns. She said the townspeople grew accustomed to this and more or less calmly went about their business.

Destiny Strikes Twice

Now that he was accumulating some potentially valuable patents, Jim decided to look for backing for his burner venture. He had already registered its name, Oil Devices Corporation, an Illinois corporation. Laurence Scudder was his first investor. He was a Chicago lawyer interested in occasionally investing in small businesses. Scudder also invented an air duct system that distributed hot air from the house and returned cool air to the furnace so he and Jim were thinking alike.

Jim also struck up an important lifelong friendship with a neighbor, Sterling Morton. Sterling was the son of Joy Morton and together they ran the Morton Salt Company. He is noted for providing the company with its famous logo and slogan, "When it rains it pours." He was also chairman of the Morkrum Company which developed the Teletype and high-speed stock ticker. It was later sold to AT&T for $30 million. Sterling had a house nearby and their children played together. He, like Jim, was interested in innovation and invention. They both spent part of their retirement in Santa Barbara, California. It is not clear if any of the Morton interests invested in Oil Devices, but I think they did.

Jim felt that it was time to move out of the Chicago area. He needed a place where he liked the weather and the people better, and where he could acquire low-cost laboratory, engineering, and manufacturing space. To do this, he revived his abilities as an aviation pioneer. By flying to points west, he might come across a new place to settle down. Eager to start this new life adventure, with two partners he helped organize an air charter company that would fly tourists over the Grand Canyon.

Destiny Strikes Twice

5

Jim Discovers Santa Fe

Jim relived the NC-4 startup as the engines coughed to life in the Ford Trimotor. He was flying it to Winslow, Arizona for the Grand Canyon sightseeing business. They would pick up and drop off passengers at the Winslow train station on the Santa Fe Railroad. This train route ran from Chicago to Los Angeles and featured western tours managed by the Fred Harvey hotels and restaurants. Their famous La Posada hotel would be built at the Winslow station two years later and still is in operation.

At this point in 1928, there were hardly any commercial flights. The upstart airlines of those days usually flew the nine-passenger Ford Trimotor. It had a cruising speed of about 100 miles per hour and a range of about 500 miles. It took 48 hours to travel from New York to Los Angeles on a trip that combined some segments by plane and some by railroad Pullman car.

The Ford Trimotor in length and wingspan was about two-thirds the size of the NC-4. Its three air-cooled radial engines generated about 1,350 horsepower—about 84% of the NC-4's four V-12 water-cooled engines which generated a total of 1,600 horsepower. Although Jim was flying a plane comparable in size to an NC-4, he had only the help of a copilot and not the vast organization of the U.S. Navy.

Ford Trimotor like the plane Jim flew to Santa Fe in 1928

They planned to land in Albuquerque to refuel and continue to Winslow the next day. As Jim and his copilot looked out of the cockpit, they didn't seem to be moving. Large mountains loomed up to the right of them, and they seemed to just stay there. "Must be a 50 miles per hour headwind," Jim said, "and our two side fuel tanks are empty. Only our middle tank has some left. Do you see anywhere we can land before Albuquerque—preferably, *right now*?" The copilot studied the chart and said, "The town of Santa Fe is about 10 minutes away 30 degrees to your right. The runway is probably packed earth but might be long enough for us." They passed over Glorieta Pass, the mountainous east gateway to Santa Fe. After a few minutes of staring intently at the ground, they saw a big arrow painted on the roof of a large building. They turned in the direction it pointed and saw a little airfield below. Jim commented to a magazine interviewer years later, "We circled and landed—just in time. Our engines sputtered and quit as we turned to taxi up to the parking line."

After arrangements were made for fueling the plane and the copilot found a hotel, Jim looked through the phone book in the coffee shop. Sure enough, his memory was right. Listed there was Bronson Cutting, a classmate from Groton. Cutting had built a career in state politics and later became one of New Mexico's U.S. senators. Unfortunately, he died seven years later in a commercial plane crash.

Cutting showed Jim around the ancient Spanish, later Mexican, later American capitol. It was and still is an enchanting place populated by single-story adobe houses and friendly people. The weather is warm and sunny year-round. Previously only known to eastern traders, the town was now being discovered by artists and tuberculosis sufferers seeking treatment in a dry climate. There was not, however, any manufacturing industry Jim could use. But the wheels of his mind were already turning.

When Jim was shown a property on Upper Canyon Road, he knew that it would be perfect for his residence, shops for the business, and stables for horses. The house was a large single-story adobe on high enough ground so it had a commanding view of Mt. Picacho. Running along the lower edge of the fields was the Santa Fe River, more properly called a creek. This supported a lot of trees including willows, cottonwoods, and poplars. On the north side of the property, by the river, was a series of small, interconnected adobe buildings. Jim saw these as perfect for his lab, shops, and office. He loved to ride, so as a priority he built a stable.

Los Vientos, the Breese house in Santa Fe, about 1930

He called his wife Marjorie back in Illinois, and she agreed to come out and see the town and property. Fortunately, there was direct train service from Chicago to Santa Fe. Her only requirement was that there be a one-room schoolhouse nearby where she could teach young children. Jim purchased the property and Marjorie, the four children, and her mother, "granny," moved in 1929.

In 1930 he renovated a building on the other side of town to be his test and limited manufacturing facility. It came to be known as the "pilot plant." His large-scale manufacturing would be done by contract manufacturers and licensees in the Midwest. The most important of these was Columbus Metal Products in Columbus, Ohio. Jim also had a sales office in Chicago. During those years he was doing business as the Oil Devices Corporation.

Fueloil & Oilheat

**A building near the house converted to test labs.
Note the high air vents for all the burners under test.**

Jim was settling into his Shangri-La and things were starting out well. His father, however, was suffering huge financial losses from the Crash of 1929. He was a widower and was forced to sell *The Orchard* mansion in Southampton and move into a two-bedroom house he built nearby. The walls and gardens had an English country look. He took an around the world trip by steamer, became a shortwave radio enthusiast, and did much of his own cooking.

James L. Breese, Sr. had charmed a beautiful young companion and toured the country with her. He decided to stop in Santa Fe. He drove up in a roadster, reminiscent of his road racing days, and he wore goggles and a duster.

There is no record of what Jim and his father talked about. But Jim didn't have much money at that time so he probably made a case for investment in Oil Devices (recall that he had been cut out of his inheritance because his wife was Catholic). Owing to his father's financial misfortune, Jim probably hadn't received much if any financial assistance.

There was, however, another matter which probably was on both of their minds. Jim's wife Marjorie was suffering from clinical depression. While this sometimes silent condition didn't seem to be a factor in her active younger years, it bothered her now. She probably had difficulty with the pressures of Jim's launching his new business and the challenges of learning about the Hispanic culture and desert environment. Furthermore, Jim worked on his inventions all his waking hours and he often took business trips to the Midwest so he didn't have much time left to be with his wife and children.

Probably with his father's support, Jim sent Marjorie to a clinic in Chicago for rest and treatment. When she returned, however, Jim still found her difficult to live with and in 1934 he divorced her.

Later that year, Jim's father died two days before Christmas and just three days into his 80th year. The Old Order had passed. Jim had no further reason or interest to be on the east coast except for a reunion with his sister and two brothers. He had to focus on pursuing his destiny in the west.

Marjorie and her daughter Ann, mother "Granny," and son Jim at her house in Ranchos de Taos, New Mexico, after her divorce.

6

The Fun Thirties

Jim Breese undoubtedly enjoyed the new home in Santa Fe with his fun family. His three daughters and son were teenagers during much of the decade of the thirties. For them, it was a new world to discover and many exciting things to do, including riding, swimming, and camping. His youngest daughter Ann wrote:

"Father hated commuting in and out of Chicago and both of my parents were glad to escape from the damp cold winters. Now he had only a three-minute walk from his house to work.

"The Martinez family lived above us in one of the adobe houses on Cerro Gordo Road; and we were fortunate to have young Andres Martinez come to work for us and take over some of the outdoor chores, which included caring for our growing animal population. It was not long before we had several horses, two sheep, a goat, and a cow. Andres also cut the lawn and alfalfa and was an expert at making adobe bricks when the house and outbuildings needed repairs. He had several children all of whom became our playmates, and their pretty, good-natured mother, Philomena, came regularly to help my mother in the house.

"Soon after we moved into our house, our Hispanic neighbors helped to name the property *"Los Vientos,"* which means "the winds" in Spanish — sort of a pun, but close enough to be a translation of "The Breeses." Coming upon a piece of our old stationery when cleaning out my desk with only *Los Vientos,* Santa Fe, New Mexico, for our address, I could not help but long for those simpler days when there was no need for a street name and number to receive mail — let alone a Zip Code!

"The family yearned for a place to swim, and we soon busied ourselves fashioning a kind of grotto by piling rocks to dam up the Santa Fe River near some Cottonwoods where it flowed through the property. We would all troop down to splash around in the cool water particularly refreshing after a long ride under that fierce New Mexican sun.

"Later, of course, a real swimming pool did materialize. A friend who was around in those days writes that Dad diverted the Santa Fe River through the pool and back into the river bed. In winter, the pool sometimes froze; and some of the older children went out curling with broomsticks and skates.

"Artist Randall Davey, who lived and kept his studio about a mile up Canyon Road from us (now a notable Audubon museum property), often came to *Los Vientos* to swim. He and Dad thought it would be a "bully" idea to install two trapezes, one at either end of the oblong pool.

Each trapeze was reached by a ladder, topped by a small platform (Author note: my father Peter Kilham, who had an ironworks in Santa Fe, built the trapeze structures.). Once on the platform, the next step was to throw out a grappling hook and pull in the trapeze. Somebody on the opposite platform would hurl out his trapeze, while at the same time leaving his platform to swing out so there would be a midair transfer leap from one trapeze to the other across the pool. Davey loved to show his trapeze artistry by including a somersault before grasping the oncoming trapeze, while, of course, all onlookers clapped enthusiastically.

Is everybody watching?

Oops! Stuck between both trapezes!

"As riding became more and more a part of our lives, the big event of the summer was the Santa Fe Horse Show. This was started by a small group of people, mainly Eastern transplants headed by Martha and Amelia White, two sisters who had come to live in Santa Fe at the same time we did. They were excellent riders and interested in promoting English style riding, with flat rather than Western saddles. They were good organizers and talked my family into having the First Annual Santa Fe Horse

Show in front of our house. Events were offered for both Western and Eastern riders. Friends of all ages would ride up Canyon road during the early summer to join the Breese girls for daily practice. There was a ring, and jumps were set up. With my mother, father, two sisters, and me all riding in these shows, the Breese family often managed to corner the market when ribbons were handed out.

"Preparations for the big day included much activity in the kitchen with the cooking of hams and roasts and pots of beans and building a temporary bar the full length of the portal. In those early years, everyone who rode in the show, as well as the audience, was invited for a buffet supper. My little brother and I were not allowed to stay up for all the festivities later in the evening, but we heard far into the night the guitar music, the singing, and the clomp of boots of cowboys dancing the polka.

Jim jumping with his daughter Frances   **On his horse Jewett**

"When the horse shows came to an end because they had grown too large for *Los Vientos*, they moved to the county fairground and our field was used for polo practice. Dad had been able to gather a few interested players, many of whom had schooled quarter horses or cow ponies into performing very well as polo ponies. There was actually a Santa Fe polo team that went as far north as Colorado Springs to play matches and often down to southern New Mexico, where young artist Peter Hurd, an avid player, provided his ranch for the matches. Sometimes I would ride along in the truck hauling the ponies. It was my job to walk them between chukkers and, of course, be a one-child cheering section for the Santa Fe team.

"The Santa Fe *Fiesta* was always the major event at the end of the summer, and it still is. It was the annual celebration staged to commemorate the Spanish reconquest of Santa Fe. In the thirties, the festivities lasted a whole week; and people wore Indian or Spanish costumes all the time. One of the most inventive and talented artists in Santa Fe in those days was Will Schuster. In 1926 he designed and built a huge, ugly, monsterlike figure called Zozobra or Old Man Gloom, an image of depression and darkness, embodying man's misfortunes. Zozobra was an awesome 60 feet or so in height, standing on a raised platform on a hill north of the plaza, and all of Santa Fe came out to watch the ritual burning of this effigy. He let out shrieks and growls and roars that could be heard for miles around, on the eve of Fiesta, he was to be burned to the ground, thus allowing the festivities to

Destiny Strikes Twice

officially begin. Part of the ceremony leading up to his being set afire in those early days was to have a group of young girls costumed in white robes perform a kind of Martha Graham-inspired dance around him, accompanied by Indian drum beats that steadily grew in intensity till one of the dancers—one year it was me!—was given a flaming torch to set fire to the monster, whose groans by now had become more deafening and were intermingled with the cheers and the honking of horns from the onlooking crowds. Old Man Gloom was dead; let the fun begin!

"An equally spectacular event during Fiesta was the history parade celebrating the re-entrance of De Vargas, who returned to Santa Fe and put down the pueblo Indian rebellion in 1692. Always a handsome young man, chosen from one of the old Spanish families and decked out in the costume of a conquistador, headed the parade on his prancing steed, followed by his soldiers in their plumed helmets and armored vests, followed by Hispanics, Indians, and Anglos, all dressed to represent different periods of Santa Fe's history. *Los Vientos* was always full of guests at Fiesta time. Often relatives made the trip west just to participate. The best Mariachi bands from Mexico would be playing all around town and in the hotels, and there was dancing in the streets both day and into the night. It was the time for giving parties and going to parties.

"We were fortunate to have at *Los Vientos* in the mid-thirties a fascinating young lady, Carmen Baca, who took over the care of my brother and me after my parents divorced. The Baca family lived directly across from us on Canyon Road. Margarita Baca, Carmen's mother, a widow, had been the New Mexico Secretary of State for years until her retirement—a tiny, vivacious woman who was half French, half Spanish. Her five children had been brought up on her husband's sheep ranch in Northern New Mexico and were schooled by a tutor brought from Spain. Their life on the ranch was well documented in a series of articles in the *New Yorker*, written by Oliver La Farge after he married Consuela, Carmen's sister. The Baca girls taught us many of the Spanish folk songs, as well as popular cowboy songs of the day. From them, we learned to do the Mexican dances, La Varsoviana, and La Raspa. We learned to speak and to read Spanish, and we even wrote and recited simple poems in Spanish. We were lucky to have Connie during those times that our mother could not be with us.

"It is important to remember that Santa Fe has been invaded many times over the centuries, and with each influx of new and different people, it will change. But underneath the newly applied glitz and ballyhoo of the 1990s, the land and sky will remain the same, and I will be drawn back again and again. I am grateful that my father was intrigued by this place and moved his family there. Those early years at *Los Vientos* will always be among my best memories."

In the early 1950s I spent some of my early teen years at grandfather Jim's place, *Los Vientos*. My mother had a house on the grounds made of adobe bricks she made years before. It was still a delightful place to grow up. I worked informally in Jim's labs, one of his employees took me on camping and fishing trips in the high mountains nearby, and I rode my horse up into the mountains.

I recall when Papa Jim, as we in the family called him, took me along on his visit to his burner pilot plant across town. Afterward, just at sundown, he took me to see the airstrip where he landed the Ford Trimotor in 1928. The town had grown around the airstrip and all that we could see was a dirt track in the weeds. Small houses, abandoned cars, and stray dogs had taken over the airport. He was silent for a long while and then said, "Well, we have a new airport now out beyond the country club, but I always think about landing here." I didn't realize it then, but Jim was preoccupied with serious personal issues.

After Jim divorced Marjorie in 1934 he married Sarah Spencer Morgan Gardner about a year later. He knew her from his college days at Princeton. She grew up in The Morgan Mansion in Princeton, daughter of Junius Spencer Morgan and Josephine Adams Perry Morgan. In 1914, a year before Jim married Marjorie, Sarah married a well-known broker and socialite in New York and Princeton, Henry Burcell Gardner. He died in 1932.

This might have been a marriage of star-crossed lovers, but family members feel that Jim had run out of money and Sarah was willing to bank him. They might have lived happily together, but issues of children and place came up. She wasn't interested in Jim's children except for NC, who she took on trips to Palm Beach and elsewhere. Jim, however, did not like Palm Beach and Sarah was losing whatever enchantment she had with Santa Fe.

Personality and culture clashes boiled over in the Great Shootout, a story passed on in Breese family lore. Sarah's daughter Sarah Morgan Gardner was married at *Los Vientos* to Sumner Rulon-Miller, know as "Ippy." Ippy, the star fullback of the Princeton University football team, did not, however, feel comfortable in Santa Fe. Ippy was miserable with the help Jim had around the house and grounds. He called them "just a bunch of Mexicans."

One day he got into a terrible shouting match because he wanted to kill Andres' chickens—"disgusting creatures." Andres Martinez (noted at the start of Ann's story above) was drinking and they dug into positions. Ippy was among the trees and Andres holed up in the tack room. Andres grabbed all of Jim's guns that were kept there—pistols, a shotgun, and a rifle.

Then the shooting started. Maybe Ippy managed to call the police. Three quickly showed up and took positions behind the terraces. When Andres wouldn't come out, the police opened fire and filled the tack room with holes. Andres hit a cop in the shoulder but he survived. Later Andres visited him in the hospital and they became friends again. Jim, through his connections at City Hall, managed to get everybody exonerated and he agreed to send Andres to alcohol treatment.

Sometime in the late 1930s while driving to Chicago, Jim picked up an attractive young hitchhiker named Irene. She was a nurse originally from Austria and seventeen years younger than him. They became lovers and when Sarah found out about this, she divorced Jim in 1939. A year later Jim married Irene. Irene Rich (given name: Anna Josefa Irine Sobczyk) was known among the children as "JLB #3" and was not well-liked by the help either.

Irene probably prodded Jim with something like, "Why are we living in this ancient place? You deserve something better!" In 1940, Jim hired the iconic Santa Fe architect John Gaw Meem to expand and modernize his house. Basically, he wanted a second story and a conservatory. The second story, a bedroom suite with a stunning view, was added with a setback from the first story below.

The conservatory was a semicircular addition to the left side of the house. It had a lot of glass bricks in its construction—a technique Jim probably remembered from his time in Chicago where they were used much more frequently. I recall large leaf tropical plants constantly sprayed with a mist. The mist would give the added benefit of humidifying the house. In Santa Fe the air is uncomfortably dry for many people.

The house design style was changed from the traditional brown adobe with round beams called vigas poking out from the upper parts of the outside walls. The new style, called territorial revival, was flat white stucco with the upper walls and flat roof joined by bricks.

The renovated *Los Vientos* in the early 1940s

A sad note to the house story is that in the mid-1940s Jim was friends with scientists working in Los Alamos on the atomic bomb. With their leader, Dr. J. Robert Oppenheimer, they would visit *Los Vientos* for parties and fun at the pool. During one party when they were in a silly mood, the usually serious scientists wrote their signatures on a wall of the house. Irene objected to this and had that bit of history painted over.

The fun thirties had come to an end and the nation was sinking into the war years. There was severe rationing for consumers and industry for food, gas, and basic materials like steel, but Oil Devices Corporation nevertheless was building a healthy business.

Destiny Strikes Twice

My mother Frances and her sister NC fishing to dramatize potholes

7

Building the Oil Burner Business

Jim Breese had many credentials as a leading engineer in combustion devices. He was the engineering officer on the NC-4 with a specialty in large engines. In 1923 Jim designed the first oil-burning boiler. He originated the first combination of hot water and oil space heating. He then went on to invent the first practical oil control valve. He was the first to build a stove around a wickless oil burner. Automatic draft control was perhaps his greatest contribution to the field of natural draft oil burners.

With this combination of engineering and technical ability, Jim could safely start his oil burner business in Santa Fe, New Mexico, a tourist haven far from any technology or industry. He would continue to have most of his manufacturing done by subcontractors and licensees in the midwest, but that strategy would not have worked without the large and ever-growing "fence" of patents he brought from Chicago and further amassed in Santa Fe

Destiny Strikes Twice

Jim's office and labs by Santa Fe artist Fremont Ellis

He remarked to *Suntrails* magazine: "Eastern markets seemed so far away, but now that we're dealing on a world market, the eastern states seem so close. That's one of the things that makes the whole situation dramatic to me. Although Santa Fe is so unindustrialized and isolated, modern communications and transportation can let it be a forerunner in a development like this." And he could only dream about the likes of the internet and FedEx. With these facilitators, I felt confident in moving my high tech international business from New Jersey to Santa Fe in 1993.

Early in business as Oil Devices Corporation, Jim developed his most important business relationship:

Columbus Metal Products in Columbus, Ohio would become his primary burner manufacturer. Most of his burners were sold to other heating equipment manufacturers. Sometimes Jim's company was paid in royalties and sometimes by straight burner sales. Columbus Metal Products also invested in Oil Devices and their president, Harold B. Donley, served as a director. For many years. Columbus Metal Products manufactured orders for Jim's customers with some totaling more than 100,000 burners per order.

Jim working on a new burner concept

Destiny Strikes Twice

In his first few years in the burner and heater business, starting in Chicago, Jim thought of selling directly to the consumer. This is a natural inclination for inventors. There are millions of consumers for anything. This must be the market to go after. Cut out the middleman! What the fortune gazing inventors tend to overlook is that there are usually two steps from the factory to the consumer: distribution through industry specialists and after them, retail through stores or, these days, online. A lot of capital is tied up in warehousing the products in distribution, and there is a lot of expensive expertise involved in promoting the products to the consumer.

Jim envisioned a wide variety of consumer products. There was the large and obvious opportunity to replace the millions of coal home heaters in use during the 1920s and 1930s. His oil burners had automatic controls and burned cleanly. They were also low in cost. So Jim saw an opportunity to sell oil heating to millions of customers at affordable prices. The products would be space heaters and floor furnaces. After a short while, he gave up the temptation to design and manufacture products to sell directly to the consumer. He made standard burners which he sold to established manufacturers of space heaters and furnaces and these companies in turn sold to the major retailers like Sears and Montgomery Ward. Today they would sell to retailers like Home Depot and Lowe's.

By 1951 five million American homes were heated by oil. Two-thirds of those installations used Breese Burners or incorporated Breese patents from which he received royalties.

Jim experimented with other consumer products but none of them proved significant for his future. He tried water heaters and water heaters integrated with furnaces but to no great success. He also looked at camping stoves, and eventually had some projects with Coleman, but nothing significant materialized there either.

Flyer for proposed Breese space heater
Height about 40 inches

Destiny Strikes Twice

By 1935 Jim through Oil Devices had sold and licensed a million burners. In 1936 alone, 161,506 heaters with his burners were sold. His sales in the 1930s were in the hundreds of thousands of dollars per year which was timely to finance private education for his four children, improve his properties, and invest in people and equipment for his growing business.

Then about 1940 Jim discovered a lucky technical break for his business. Using a certain type of his patented fuel valve for his heaters, they could burn gasoline, kerosene, aviation fuel, or the conventional fuel oil with no other changes to the equipment. This was perfect for the U.S. Army. A burner in the field would be much more useful if it could burn any kind of fuel available in the fuel dump. No other burners on the market had this flexibility.

His son Jim Breese wrote to me about this development: "Of course, there was a lot of official resistance to the idea of burning gasoline. The Corps of Engineers types at Ft. Belvoir visualized cataclysmic explosions, and a colonel from there flew out to discuss the matter with Dad. They sat next to a Jerry can full of gasoline while Dad tried to explain that liquid gasoline doesn't explode. Then he lit a cigarette and tossed the burning match into the open can. The colonel, white as a ghost, hit the dirt. But, of course, the match fizzled, and the point was made."

The Breese burner was a huge improvement over the coal-burning tent heaters. Jim sold the Army "drop-in" burners to retrofit the coal burners in tent stoves, new complete tent heaters, and "immersion" heaters used to heat the water in large steel barrels. Veterans probably will recall rinsing their mess kits in the hot water and shaving with hot water poured into their helmets.

Jim directly or through licensees sold several hundred thousand burners to the Army during World War II. His contract prices covered burner costs with a profit but did not include royalties on the patents used. Many suppliers to the U.S. Government did this in the name of supporting the war effort, and Jim may also have been thankful to the Navy for including him on the NC-4 flight.

By 1947 Jim had about 30 employees in Santa Fe and a salesman in Chicago. 15-20 of his employees worked in the Santa Fe pilot plant. While he was by no means Santa Fe's largest employer, he was certainly the city's largest technology and manufacturing concern. With this prominence, the horse shows, and his famous swimming pool, he had attracted a large circle of artists, writers, and scientists.

One great source of irritation to Jim was the Valjean patents. Ben Valjean was an engineer who assigned his patents to the Motor Wheel Corporation in Lansing, Michigan. They controlled about a third of wheel production in the United States, and their interest in oil burner technology is not clear. The first patent, issued in 1931, was for an oil burner pilot and valve. It was a simple mechanism for igniting the pot type burners like the Breese design. It premixed the ignition oil with air to facilitate the main combustion after that in the burner pot.

The second Valjean patent, issued in 1937, was a method to assure complete combustion of the fuel in the burner. It used a circular baffle or flat ring about halfway up the burner wall. In 1939 a patent was issued to Jim for essentially an improvement on the 1937 Valjean patent although Jim's burner design had been using a similar principle since 1927. If this sounds confusing, it was also confusing to engineers and patent attorneys as we shall see in the next chapter. Eventually, Valjean assigned use of his patents to Jim. This resulted in major royalty contributions to Jim's company.

In 1947 Jim reorganized his company. He changed the name from Oil Devices Corporation to Breese Burners, Inc. Breese Burners published an annual report for 1950 showing sales of $994,049 and net income of $74,963 ($10,716,000 and $808,101 in 2020 dollars). The number of shareholders doubled from the previous year to 37 and the number of employees also doubled to 52. The shipments of products increased from 132,501 units in 1949 to 267,148 units in 1950.

Sales to the government in 1950 were $561,552 or 56% of total sales. Defense work was rapidly increasing due to the outbreak of the Korean War. This will have major consequences for Jim.

From 1943 to 1946 Jim was in command of the New Mexico Wing of the national Civil Air Patrol (CAP) with the army rank of lieutenant colonel. He was, of course, a nationally recognized military aviator and acknowledged leader, but I also think it was a good escape from the pressures of his business and home. The CAP was formed to search for enemy activities in the homeland and to conduct search and rescue missions.

Lt. Colonel James L. Breese, New Mexico CAP, 1943-1945

Destiny Strikes Twice

In December 1943, Jim received a letter from Endicott Peabody, Headmaster, Groton School where he had prepared for college:

> I am so glad to get that tome which contains the description of fifty "Oil Device Patents." In the early days I had a feeling, perhaps it was only a hope, that my knowledge might be superior to yours, —now I surrender. I am indeed amazed at the genius which proclaims itself in every one of the fifty inventions which are described in the book which you have been good enough to send me. After giving it to members of my family I shall lend it, for a time, to the masters in our science department. They will gather from it a higher respect for Groton learning, if I may put it so, than they have gathered in the years during which they have been doing good service as substitutes for some of our regulars. It may seem a little embarrassing, as you say, to decide upon our respective titles. I confess that I feel inclined to take off my hat to an outstanding inventor and to address him as "Sir." Come up some time and talk to our boys and masters concerning the important work that you are doing. I should like to present to them one who has had his early scientific training at Groton.

I found this letter among Jim's collection of special letters and with a copy taped to the inside front cover of his patent book. I think he was very touched that someone of importance admired him as an inventor.

8

Partners and Successors

At some point, every business founder must plan for and carry out succession. The traditional method is to bring in a family member and if that doesn't seem promising, often the business is sold.

Jim knew my father as his son in law for a few years before giving him a few projects at Oil Devices in the mid-1930s. The first project was designing the enclosures of some of the space heaters they planned to market directly but never did. One of Peter Kilham's designs is shown in Chapter 7. He also devised ways for fabricating the sheet metal parts of the burners. This started my father's development of patented metal bending machinery which became his business for many years.

My father told me Jim was loaded with charm and was a great salesman but found Jim impossible to work for. If anyone else thought of a technical idea, Jim put it down. The emerging friction was calmed when my father moved east for the World War II manufacturing effort and never returned to Santa Fe. Nevertheless, he and Jim were friends and correspondents for their lifetimes.

Destiny Strikes Twice

Jim knew about everyone in the heating industry, and he thought he found the perfect person to help him run the business. In 1938 he convinced Bruce Hayter, a well-known heating equipment engineer, to come west to be the key man in his growing business. He would start as chief engineer of Oil Devices (name changed to Breese Burners in 1947) and would then become the general manager of the business and a partner in the company.

Hayter was a senior official of the American Radiator Company in New York and the chief engineer of their research labs in Buffalo. American Radiator was the foremost manufacturer of steam radiators for heating buildings.

Hayter started well in engineering at Oil Devices, filing his first patent in 1939. He continued to patent technical breakthroughs through 1950, totaling 50 patents for Oil Devices. As a manager, however, he couldn't seem to find much to do. Probably Jim didn't delegate much management to Hayter. Certainly, there wasn't complete communication between him and Jim. Employees didn't like him and preferred to change the subject when I asked. In short, Hayter doesn't seem to have been the right one to succeed Jim in his business.

There was also culture shock, at least for Bruce's wife. Although they settled in one of Santa Fe's best neighborhoods at that time, she was not charmed. They had moved from Birchall Drive, the most exclusive neighborhood in Scarsdale, a wealthy suburb of New

York. She was not one of the enthusiasts from the east who were captivated by Santa Fe's Indian and cowboy culture.

The inevitable divorce happened and Hayter resigned from Breese Burners around 1950. He started a competing company across town and filed seven patents for them. Nevertheless, it soon failed. Bruce moved to a charming town in the hills of Mexico favored by expatriates. Jim's daughter Ann ran into him by chance when she was vacationing in Mexico and reported that he was at last contented with life.

In the mid-1950s Jim was suffering from declining health and his consumer market was declining due to competition from other technologies like natural gas. His military markets were up and down, depending on the latest conflict, and his last project for them was snow melters for providing water to Arctic troops. Maybe his son, my Uncle Jim, could rebuild the business somehow.

When we were relaxed over a beer in San Francisco, I asked Uncle Jim if his father had ever talked to him about coming into the business. My uncle was, after all, an accomplished engineer in heating and air conditioning. Jim answered, "Dad gave me a desk and but never told me what he wanted me to do. For two weeks I sat there every day, waiting for something to happen. So I left." I could hardly believe what I was hearing. The scion of a noted New Mexico inventor and bon vivant was recalling his lost future when he, the only son of Jim Breese, could build a bigger and better Breese Burners, Inc. With the cultural and material inheritance he was born into, how could the young Jim Breese back away from the call of fun and fortune?

James L. Breese III

James L. Breese III was a sensitive young man and not athletic like his sisters. He was not aggressive or interested in leadership. Nevertheless, when he became draft age, he joined the Air Force near the end of World War II. He was a talented writer and was assigned to report on the

Nuremberg Trials. With all the lurid descriptions of the Nazi death camps during the trials, this must have been traumatizing for him.

But things got worse. After returning to New Mexico, he often stayed with his mother in Albuquerque where he had begun studies at the University of New Mexico. She was divorced and had married an opportunistic drifter who specialized in marrying wealthy women. In 1948, at the impressionable age of 21, Jim found her dead, hanging in her closet. His mother, whom I never knew, was idealistic and had bouts of severe depression, but to this day there is no agreement about the cause of her death. Young Jim must have been traumatized.

Three years later Jim graduated in engineering from the university. He started his career working at the Los Alamos Laboratories on the Maniac, one of the first electronic digital computers. Things seemed to be going well but Jim was not an organization man who would be permanently happy in a bureaucratic government research lab. So when his father called him to come down off The Hill (as Los Alamos was and still is known) and take a look at Breese Burners, Jim agreed. I can only speculate about what his father was thinking.

The proprietary technology of the Breese oil burners was shifting from the burner designs to electronic controls. Indeed, when Breese Burners, Inc. was sold several years later, the sale was to a major industrial controls company. Young Jim, the electronics whiz, would have had just the technical background to take the family business to the next level.

Destiny Strikes Twice

When the elder Breese gave his son a desk in the company, maybe he was hoping to see where his son's curiosity and initiative would lead him. Perhaps he'd grab a file of sales correspondence and go visit customers — a traditional way for promising managers to start in small businesses. Maybe he'd look at burners in the development lab and see if he had some promising ideas for improving burner controls designs. Then he could develop a new patent portfolio for the company.

Maybe he was afraid of or awed by his dad. Undoubtedly, young Jim must have felt that his father, a womanizer who married four times, was at least partially guilty of his mother's untimely death. All I know is that according to Uncle Jim, he just sat there at his desk, waiting for something to happen.

In 1959 Uncle Jim moved with his new young wife and baby daughter to San Francisco, her home town. They settled in a small romantic apartment on Telegraph Hill. Jim set up an engineering consulting business in a low rent district near the waterfront. He specialized in heating and air conditioning systems design, and his clients included international airports, high tech silicon valley firms, and hotels. He was compensated by commissions from the sales of manufacturers' equipment that he represented.

A few years later his wife divorced him and Uncle Jim started publishing a heating and air conditioning magazine funded by the meager inheritance he had just received. He told me that his father did not care for his wife and that chill contributed to their breakup.

Meanwhile, Uncle Jim's editorial assistant for his magazine was a novelist and she later became the love of his life.

Providing continuity and growth for family businesses in the later stages of their development is a challenge that many entrepreneurs face, especially for high-tech businesses. In many cases, family stresses and disagreements confound otherwise partnerships and succession agreements. This is why many family technology businesses wind up being sold to corporate buyers and this was the path Jim was forced to take.

Destiny Strikes Twice

9

Troubles With Patents

No matter how hard you try to protect your patents, something goes wrong. Someone will try to go around you or invalidate your patents. You must be alert to the attack before it goes too far. Jim discovered the attack on his realm in a very unexpected way.

His lawyer, John T. "Jack" Watson, was also the commanding officer of the 726th Anti-Aircraft Battalion of New Mexico National Guard. Lieutenant Colonel Watson was leading his outfit in Pusan, South Korea, site of the famous battle of Chosin Reservoir in 1950. This was one of the major battles of the Korean War.

Jack was heating his shaving and wash water in his helmet placed upside down on a space heater. This was in the common room of the officers' quarters in an old college the army had taken over. He did a double-take and thought, "That looks exactly like one of Jim's burners but the label says 'Made in Japan.' I better write to him about this on the double."

Jim searched all his records about any sales or licenses issued to any Japanese company, or for that matter any company, during the time of the Korean War. Nothing was to be found and, according to Jack, the space heater would have been installed by the U.S. Army. The Koreans used a lot of space heaters but they all burned charcoal.

Jim had several of his people investigate purchases by the Army of the Breese-designed burners in the latter part of 1950. Those burners were indeed manufactured free of the royalties legally owed to Breese Burners. His team also acquired some of the illegal burners to study and show that they were copies of the Breese design.

He was in a sad dilemma. Jim really didn't want to sue the U.S. Government who had been so important in furthering his career. On the other hand, they had taken advantage of his patriotic offer for their purchases of his burners royalty-free during World War II. He was witnessing the grim realities of human nature. But there would be more to come.

Jim, through Breese Burners, decided that this matter had to be cleared up, cost what it may. He contacted a Washington lawyer he knew from school days, Huston Thompson, to represent him. Thompson was well-known in Washington for his political savvy and connections in all branches of government. In 1954 he opened the case *Breese Burners v. United States*. Representing the United States were Assistant Attorney General Warren E. Burger and Bernard Wohlfert. Burger went on to be chief justice of the Supreme Court. There were big guns in this battle.

The case history begins in 1942 when the Army requested Oil Devices (later known as Breese Burners) to convert its wood and coal burning tent stoves into oil-

Destiny Strikes Twice

burning stoves that would not emit smoke. After research on this application, Jim engineered his patented burner to fit into a standard Army wood-burning stove. The design included modifications to prevent it from smoking at pilot heat, intermediate heat, or high heat. The Army tested the stove and notified Breese Burners that it would be awarded a contract for 74,620 of these burners. The contract would include the condition that the award would be royalty-free for purchases during "the duration of the war (WW II) and six months thereafter."

The Army used this contract clause to justify using the Breese patents royalty-free during the Korean War because The Peace Treaty with Japan was not signed until September 8, 1951. The surrender of Japan, however, was on September 2, 1945, and Breese claimed that the royalty-free agreement should expire six months after that date.

In the first session of this case, the court decided in favor of Breese Burners. Having been deprived of its defense that it had a license to use the Breese patents, the Army then said the patents were invalid! They were really using the Valjean patent which preceded the Breese patent.

Breese Burners said that the Army burner was an adaption of a previously patented Breese burner which was an improvement over the Valjean patent (discussed in Chapter 7). The court record showed that claimed differences in the Breese patent were an improvement over the Valjean patent. The commissioner of the court, who was present at the tests of burners constructed according to both patents, found that the Breese burner was in fact more efficient when burning at low or pilot heat than the Valjean burner.

In 1954 the court said, "It is held that the (Breese) patents are valid and that the defendant is liable for infringement. A royalty charge of 25 cents per unit is fixed by the court and entered for the plaintiff for $146, 580.50 ($1,418,894 in 2020 dollars) based on the purchase by the Army of 586,322 units from persons other than the plaintiff. Interest (4%) is allowed as set forth in the conclusion of the law."

A cheer probably could be heard from Santa Fe to the Washington courtroom. After the excitement died down and the accounting was completed, it was found that legal and other costs consumed much of the royalty award. Nevertheless, Jim felt vindicated.

But there was another set of demoralizing costs that probably deeply troubled Jim. Two of Jim's senior staff, an accountant and Bruce Hayter whose testimony would be helpful at the trial, refused to testify unless they received a percentage of the award to Breese Burners. They both made valuable contributions to the company over the years, and Jim found it difficult to recruit professionals to move to Santa Fe. Nevertheless, he must have found their extra trial payment which ran into the thousands of dollars indecent when he was potentially in trouble.

Gun-type oil burners and gas burners were continually eating into the pot type oil burner market. Now that it was clear that the pot type oil burner business had seen its best days, Jim became more of a tinkerer-inventor. He retained the minimum number of employees to meet customer orders, and he worked on several inventions including the occasional burner improvement.

Jim was intrigued by adapting the aircraft pilot's seat to automobiles for safety reasons. It had waist and shoulder seat belts that were not available in cars in those days and it had better back and neck support. He always had the aircraft pilot's seats in his cars.

He invented a system for starting a car at night when the temperature dropped close to the non-starting point. This could happen several times at night so the car would be certain to start in the morning. The alternate, which is still used in very cold climate areas today, is an electric engine block heater.

His final invention which seemed to have some commercial application was an acetylene welding torch lighter. Some ex-employees tried to commercialize it after he sold the business, but they couldn't build a business around it.

People who knew him said that Jim had mentally retired after he received the government royalty payments in the mid-1950s. He would not sell the business until 1958, but by that time there wasn't much left except a few active patents. He was suffering from emphysema and cardiovascular disease and was building a new life with a new wife.

J. Hobson Bass

Jim loved to play the drums, especially marches

10

The *Loke*

In 1942 Jim bought a magnificent eight-meter sloop he called the *Loke*. It was built in Germany in 1929 and its racing hull presented a most serene look on the water. He kept it in Lake Mead, the large expanse of water behind Hoover Dam that is fed by the awe-inspiring Colorado River. This was perfect for Jim. He loved the adventure of riverine beaches, dramatic canyons, and remains of lost Indian settlements.

Jim wrote to his daughter NC in 1944:

"We stayed on the *Loke* until Friday afternoon when we took off for Las Vegas to show Henry the sights of the fun-filled gambling town. The principal excitement we had after you left us was Thursday night when we decided to put in for the night in one of those dark, rocky coves in Boulder canyon. We sailed across the Virgin Basin by moonlight but the moon went down just as we got into the Canyon and we had to find our way in the pitch dark. This was pretty exciting but we made our way into a dark and

Exploring on the *Loke*

eerie cove. During the night the water level of the lake dropped several inches and the next morning we found ourselves pretty securely on top of a rock. It took us until almost noon the next day to get the *Loke* off. However, this was finally accomplished with no serious damage and after a good sail across the lake we arrived at the dock Friday afternoon."

Jim's sense of adventure was stirred. Not as much as the NC-4 flight perhaps, but in a way he could share with family and friends. This was adventure he would not have found in New Mexico lakes. He had accumulated earnings from the World War II Army purchases and he could apply these to buying and maintaining the yacht. I think he also enjoyed the mechanical challenges of maintaining the engine and other equipment. One scientific toy he bought was a precise telescope for astronomical observations. He tried navigating with it on the *Loke*. And it was an ideal environment to get his mind off all the strains in Santa Fe.

Surprisingly, Jim usually did not fly his Piper Cub light plane from Santa Fe to the Boulder City airport, just a few miles from the marina of the *Loke*. The drive took all day and cars didn't have air conditioning in those days. Sometimes he tested experimental car air conditioners on those sweltering trips. Also, it was nice to have a car for visiting Las Vegas and to make trips to Los Angeles, only a few hours away, which he did increasingly often.

His children told me they had a lot of fun and adventures on the boat. A letter from his daughter NC to my mother in the early 1950s shows an example of a memorable time they had. It said in part:

"We had an awful lot of rain which you would think would ruin everything but actually it was as much fun as ever. We did more campfire cooking on sandy beaches than any other trip and I liked it a lot better. We had Nelson Jay along who can really do things to steaks over charcoal. Ann (her sister), Dad and myself and then later Pitch (her husband) made up our crew. It was quite exciting as we didn't know for sure if Pitch could get away or if the weather would stay decent enough for him to get down flying Dad's Piper Cub. We planned a rendezvous in the upper lake on one of two days—no special place just wherever he found us.

"We really didn't think he would show up, but on Monday afternoon we suddenly saw this little plane almost upon us. We didn't even hear the engine approaching. He circled several times while we all waved madly and then dropped a stick with a message. But the paper came loose and flew off while poor Nelson was madly rowing after it in the dinghy. He then fashioned another message and dropped it. A great wind sprang up of course about that time and there was Dad trying to maneuver the boat with both mainsail and jib up while Ann and I flung ourselves out into space trying to pick up the second piece of wood. Finally chaos was calmed and we read the horrifying message: 'Can land on mesa.' Also, it said to wave a towel if okay. We all debated shortly and then waved him back to Boulder (the airstrip at Boulder City) and to whatever method he could get back to the boat. The weather being as bad as it was and any mesa a fur piece to hoof it, we thought it would be safer.

"Poor Pitch had to spend that night in Boulder and then finally talked someone into running him up to us the next day in a fast speed boat. We only found out then that the engine came very near to conking on him when he throttled back to buzz us. He had two other semiforced landings on the way down. Just what would happen when a little plane plops on water I don't know. We could have fished him out alright but it would have been the end of Dad's toy."

My mother's story which she also illustrated shows the dangers and delights of exploring the upstream canyons:

"We sailed across the lake, and in what appeared to be an impenetrable mountain wall we found a narrow canyon which we maneuvered into and pulled up for the night near a strip of sandy beach. Our boat, which seemed big enough when we were on it, seemed dwarfed completely by the towering cliffs on all sides. We cooked and ate supper on board, then built a good fire on the beach and ate popcorn and played a little portable radio till late that night.

"The next day, as luck would have it, a good wind came up and we had an exciting sail through the canyon, tacking one way and then the other sometimes missing the canyon wall by a few feet. Harry ran around doing things with the sails and watching for rocks, and I steered. We explored a quiet inlet where the water is very clear and we could see the rocks below waiting for us and it was fun outwitting them. We went ashore and had several hours climbing the mountains, swimming, and just lazing in the hot sun. We saw lots of wild burro tracks but no actual wildlife except for a few birds.

"One thing that is most impressive about that country is the extreme quiet. When the water was still, the only sounds would be the rare motorboat or an occasional squawk of a bird. That and never knowing what time it was, were two of the big features of the trip. Also the calmness and slowness of sailing, such a relief after all our usual hustle and bustle."

Destiny Strikes Twice

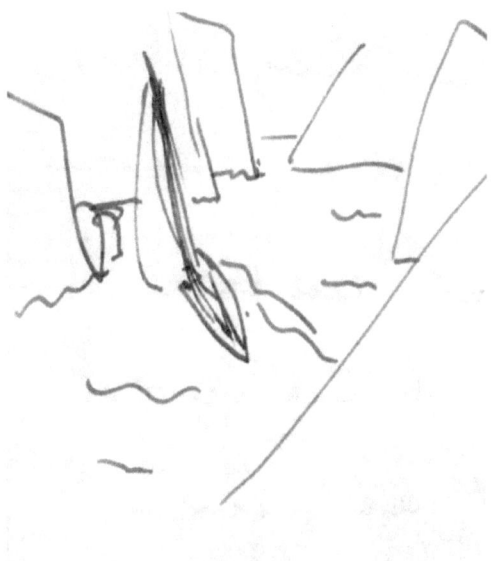

Jim's fourth and last wife Florence had several friends from Beverly Hills whom she invited to sailing excursions on the *Loke*. Several of them expressed interest in buying the boat because it stimulated them to get back into sailing again. Jim preferred to keep the *Loke*, even in his advancing years. He did make offers for them to charter the vessel, and the charters probably defrayed the considerable maintenance costs.

The comings and goings of the "Hollywood Types," as his daughter Ann called them, led to an amusing story that she loved to tell.

"One day John Wayne appeared, apparently unannounced. He thought his advance man had alerted

Jim on the Loke

Dad, but Dad, typically hospitable, welcomed him aboard. The old cowboy clomped across the decks in his muddy cowboy boots, peering at the boat's features.

"Then he formally introduced himself and allowed as how he might like to buy the Loke. They tossed that idea around for a while and then Wayne asked, 'Where's the bathroom (the "head" for you landlubbers), and while I'm down there, can I help myself to a cup of coffee in the kitchen (the "galley").' John Wayne was actually a nice guy and trying to be pleasant, but he was losing interest and departed into the sunset."

Destiny Strikes Twice

Sometime in the 1950s when he was in his late 60s, Jim suffered a heart attack while scrambling up the embankment from his dinghy to the parking lot. He was rushed to a hospital and saved but he could see that this part of his life was drawing to a close.

Jim sold the *Loke* and looked west to California. This would be his last life phase.

11

Florence

Halfway up the walk of 100 yards from his office to his home, Jim would often sit under a shady and accommodating willow tree. It was next to a trickling stream that ran down a slope and disappeared into the ground. He designed this shady area to exactly suit himself for rest and reflection. When there, Jim couldn't help thinking back to the NC-4 flight. Those were his days of great risk and great glory. He was happy then.

Jim needed the rest stop because in the early 1950s his breathing became labored. The causes were emphysema and arteriosclerosis. Contemplation satisfied his need to reflect on the triumphs and setbacks life had dealt him and to think about his future. Everyone knew not to interrupt Jim in his sanctuary.

The days of driving to Chicago and the *Loki* were coming to a close. As Jim told *Sun Trails* magazine in 1954, "My two sports were polo and flying, but I've sold my plane and I've turned my pasture into a pond. Just like an old cowpuncher, there came a day when I could ride no longer."

Jim should have been proud that he invented an important product that warmed millions of people and his company employed dozens of people in sleepy Santa Fe. He had a family with four fine children who loved him dearly. But something was missing.

My father who knew him well said deep down Jim was very unhappy. He had too many problems and always seemed to be creating more problems.

Jim's three past marriages had not gone well. His first wife Marjorie had committed suicide after a long battle with clinical depression. Even though she remarried, she had said that Jim was the love of her life. Then he married Sarah, a wealthy socialite whom he knew from college days. He was not comfortable with her second hometown of Palm Beach, and she gave up on him after discovering he was having an affair with a hitchhiker he picked up. He married Irene the hitchhiker when he was 56.

Irene had no social credentials, but she was young and pretty. Although like Sarah she was not popular with his children, she and Jim apparently lived happily together for a few years. Then tensions arose and festered until they went to court. The judge decreed that a fence be built through the middle of the house from front to back so each could enjoy his or her own bedroom, dining room, and porch. I've never heard of such a thing but I know it's true because I talked to the workman who built the fence.

In 1947 their marriage was annulled and Irene was awarded Jim's rural retreat in Glorieta, a settlement about twenty miles from Santa Fe. The property occupied dozens of verdant acres nestled in a valley at the foot of the mountains. It included a farm where Jim raised organic fruits and vegetables. Later Irene sold the property to the Baptist Church which built a retreat center and summer camp there.

Irene was also awarded real estate they owned in Golden Beach, Florida, north of Miami. They planned to build and maybe retire there. Irene was previously from Florida but Jim hadn't shown much enthusiasm for Florida beachside living. Probably this Florida real estate purchase was her idea.

Jim tried one more time to find a wife who would be mutually compatible and loving. I don't know how he found her or how long he knew her, but shortly after his divorce from Irene, he married Florence Welch Wagner in 1948. They would remain happily married until Jim died.

Florence started in Kansas as a newspaperwoman. She moved to Los Angeles in the early 1900s where she married Rob Wagner, a portrait painter and writer. Later he became the owner and publisher of *Script*, a weekly literary film magazine in Beverly Hills. Florence became the magazine's business manager and a columnist. *Script* became noted for its courageous social stands and for its notable contributors including Walt Disney, Ray Bradbury, Charlie Chaplin, and Salvador Dali.

When Rob Wagner died in 1942, Florence became the owner and publisher of *Script*, and she sold it in 1947, a year before she married Jim. The magazine was going downhill and it's doubtful that she profited greatly from the sale. She also received writing credits for three films.

Destiny Strikes Twice

**Florence Wagner Breese and James L. Breese
Late 1940s**

Florence had become a well-known personality in the Beverly Hills social circuit and Jim enjoyed meeting her friends. This was near the end of Hollywood's Golden Age so there were lots of interesting people and projects instead of the intense corporate culture endemic in the film industry today. Florence and Jim often stayed in a waterfront motel in nearby Santa Monica with glorious views of the sunsets on the ocean's far horizon.

Back in Santa Fe, a close friend of the family described Florence as "good-looking and an independent thinker. She even tried removing Jim's gas burner called 'the eternal flame' in the living room fireplace. She was very nice to him and tried to keep him alive during his lingering death from belabored breathing." Florence was stiff with his children and employees so the festive atmosphere of *Los Vientos* withered away.

After Jim and Florence had left Santa Fe for good, hundreds of hypodermic needles were found in the bedroom suite. I believe Florence, acting as Jim's primary care-giver, had narcotics sent from her sources in free-thinking Hollywood. These would be considered legitimate pain-killers today because they stopped his constant pain without him becoming a serious addict.

Jim's last faithful general assistant employee was Benny Martinez. He recalled to me that almost every day he would visit Jim in his bedroom or on the porch and Jim would describe and sketch a new technical idea — usually a burner improvement. Benny would bring the sketch to the remaining engineer to be redrawn and a model made and tested. Then Benny would return to Jim with the test results and this cycle would start again. It was during these visits that Benny discovered Jim's drug habit and he thought that Florence was on drugs as well.

During Florence's early days in Santa Fe, she and Jim decided to improve the appearance of *Los Vientos* and the grounds. In the words of the architectural historian who was reviewing the property's history years later, "The Breese residence took on an estate-like appearance during this period. It was approached by a curving tree-lined drive and included a wide expanse of manicured lawn peppered with ponds and pools, hedges and flowerbeds. It was an appropriate setting for an oil burner magnate and his Beverly Hills wife, though hardly in the same league as his father's Hamptons mansion."

The final version of *Los Vientos*

Larry Kilham

Chapter 12

The Last Flight

In 1956 Jim flew to Europe on TWA to relive his NC-4 Navy flight with a modern plane. The record hasn't survived about the departure and destination cities, but given TWA's timetables at the time, and Jim's likely interest in duplicating his earlier itinerary, it was probably New York to London. His flight would have been on a Lockheed Constellation with its iconic triple tail and dolphin-shaped fuselage. It was powered by four 18-cylinder air-cooled engines generating 3,250 horsepower. If there were bad headwinds or other problems, this transatlantic flight would stop in Gander, Newfoundland, or Shannon, Ireland for fuel and maintenance. There was a crew of five and typically 62-95 passengers.

In many ways, the "Connie" would have reminded Jim of the NC-4. The engines made puffs of smoke when they started, were very noisy with a deep throbbing sound when warming up, and flames and sparks streamed out when taking off (I remember this well having taken the Connie on Eastern's New York-Boston shuttle many times

Destiny Strikes Twice

Lockheed Martin

Jim's last flight

in the late 1960s). The wingspan at 126 feet was identical to the NC-4 but at 116 feet was about twice its length. The Connie at 137,500 pounds had over four times the maximum take-off weight of the NC-4 and at 377 miles per hour flight speed over four times faster.

If Jim had waited a few more years, he probably would have taken a jet, like the iconic Boeing 707. Although a completely different experience from the NC-4 or Connie, it would have fascinated him with its next-generation design. In any case, there's no record of Jim's thoughts or experiences on his flight.

Larry Kilham

His daughter Frances wrote this summary for him:

HAIL TO THEE, BLYTHE SPIRIT

Here's to Jim Breese
Who first flew the seas
In the biplane NC-4.
The going was rough,
And things were tough,
But they got to the opposite shore.

That was long ago.
Now he says I'll go
And give it a try once more.

Things aren't the same
Since we flew to fame
In the good old NC-4.

In seats so soft
You are borne aloft
You don't notice the roughest breeze.

The martinis are dry
As you sail through the sky
And the hostesses aim to please.

The food is fine
And served with wine,
And he who has done it before

Destiny Strikes Twice

Says look what they've done
Since trip number one.
It's not like the old NC-4!

In 1958 Jim decided to throw in the towel. Breese burners had only a handful of employees and he was having trouble paying them. New technologies were pushing his pot type oil burner for space heaters off center stage. One competitor was the gun-type oil burner where the oil is atomized just before ignition as it is sprayed from a nozzle. The other competitor was gas heat which took off when natural gas was piped to millions of homes. Safe ignition and automatic control mechanisms for gas heaters helped its acceptance.

He sold almost all of the assets of Breese Burners to Controls Company of America in the Chicago area. They bought his patents, fuel control and ignition technology, and an agreement that he would assign future patents to them. He had about 130 patents at that time and eight applications that became patents. These were all assigned to Controls Company of America. Jim was an inventor to the end!

His sales manager, Richard Van Tubergen, and a mechanic, Charles Mueller, bought the Breese Burners name and several small products including an acetylene torch lighter. This new venture failed and was never fully paid for.

Jim's income had dwindled to proceeds from his successful patent suit against the U.S. Government. This could have been more than $50,000 after legal fees and other trial costs.

After his death, his laboratories and offices became the classrooms and offices for the newly-formed Santa Fe Preparatory School. Later, continuing in the education role, the buildings were the stand-in for Santa Fe high school in the local coming-of-age movie *Red Sky at Morning* starring Desi Arnaz, Jr. and Claire Bloom.

In late 1958 Jim and Florence moved to a rented house in Santa Barbara, California. It was in the exclusive Hope Ranch enclave on Marina Drive. This gave them a spectacular view of the Pacific from a high clifftop setting. They both wanted to finally settle down away from all his Santa Fe troubles, and they hoped his cardiopulmonary problems would improve in the lower altitude with a humid climate.

Florence's husband had died years earlier in Santa Barbara, and undoubtedly there were family matters to attend to. Jim probably enjoyed reminiscing about inventions and the Chicago days with his old friend Sterling Morton (discussed in chapter four) who retired with his wife in an elegant country home in Montecito a few miles away.

This move looked promising, but they began to see that Jim's ever-worsening condition would require the essentially constant attention of top heart specialists. In early January 1959, Jim and Florence moved to an apartment in La Jolla, California, just north of San Diego.

Destiny Strikes Twice

For the next three months, he was treated at the Scripps Clinic in La Jolla.

**Last photo of Jim
Santa Barbara, January 1959**

On the afternoon of April 1, 1959, Jim took a walk on the La Jolla beach—one of his remaining joys in life. Suddenly he began gasping for breath and collapsed on the beach. He was rushed to the Scripps Clinic but died there of pulmonary edema and arteriosclerotic heart disease. He was 73. Florence died 12 years later in La Jolla.

The New Mexican, Santa Fe's newspaper, commented in their obituary: "All Santa Fe can justly feel sorry that Jim Breese is dead. All Santa Fe can be glad that he came to live with us."

Destiny Strikes Twice

Epilogue

Times of insight and creativity come and go with the ebb and flow of unexploited knowledge and with society's sense of urgency for new solutions. Jim Breese came along during such a flow. He witnessed the introduction of automobiles, radios, washing machines, and penicillin. He dared to be the engineer on the first transatlantic flight. He brought low-cost and clean heating for people of all incomes.

Now we take for granted more recent inventions and developments including the internet, AI, cell phones, and self-driving electric cars. But there seem to be insurmountable challenges like climate change and devasting environmental destruction. There is an apocalyptic sense of the world running out of time. Many people feel a sense of "Why bother?"

People must see that the whole universe is available to them and that creativity has never been more important than now. Children should realize that there is an infinite future for them. Society's failure is a failure to give them hope and encouragement.

Now is the time for the men and women who dream of things that never were. Their dreams are the starting points in great creations. The positive emotions of the challenge will cause the complexity and depth of the world's problems to fade away. The one catch is that their dreams will have to answer to unmet realities.

It is time to turn America and the whole world into a nation of creators and inventors again and for the whole world to work together to deal with the many challenges and opportunities that are upon us. From garage inventors to multinational corporations we must make a fresh effort at creativity and innovation including using the vast new resources of the internet and the computer clouds. America and the whole world need to become more creative in all endeavors.

Jim Breese would heartily agree. He would hope that he has set a good example and has given entrepreneurs insights and knowledge they didn't have before.

Destiny Strikes Twice

Acknowledgments

Special thanks go to James L. Breese's grandson Mark Breese Sink who tirelessly mined his cache of family letters and photos as sources for this book. I am also grateful to Jennifer and Joe Freeman and to my business professor friend Dr. Brad Zehner for reviewing the manuscript.

I have had many engaging sources but they have all passed on: The three Breese daughters who contributed many memories were my mother Frances Breese Forbes, Mary "NC" Breese Jay, and Ann Breese White. Their brother Jim contributed many recollections and documentary materials over the years. My father, Peter Kilham, a prolific inventor and entrepreneur, was a fountain of information. He worked with his father in law, Jim Breese, for about ten years and they corresponded about business and engineering thereafter. I had many fruitful and amusing conversations with the Breese family friend and legal counsel, John "Jack" T. Watson.

Through it all, my wife Betsy has been my editor, rock-solid advisor, and cheerleader.

Larry Kilham

References

For all the chapters references are primarily:
Breese, James L., Jr. Unpublished letters, interview notes, clippings, drawings, and photographs. This is mostly the author's collection.

To find patents, go to patents.google.com and search by inventor's name and/or patent number.

Chapters 1-3

Miller, Frances (Breese). *"Tanty,"* a trilogy of Jim Breese's sister's life and related side histories of her family. Privately published, The Sandbox Press, Sag Harbor, NY, 1979, 1980, 1981.

Sink, Mark. "James L. Breese and the Carbon Studio NYC," gallerysink.com/marksink/Breese_Essay_Sink.html.

Chapters 2-3

See an excellent description of the NC-4 by Capt. Tim Kinsella, Commanding Officer, Naval Air Station,

Pensacola, Florida, April 2020. The video was originally on Facebook under the "National Naval Aviation Museum" page. It can be found at https://bit.ly/3esLVGN .

For fascinating motion picture footage of NC-4 history and the actual flight, search: YouTube U.S. Navy the Great Flight of the NC-4 Flying Boat 25464. (Its copyright does not permit a direct URL link.)

Curtiss Aeroplane and Motor Corporation. *The Flight Across the Atlantic*, New York, 1919.

Smith, Richard K. *First Across! The U.S. Navy's Transatlantic Flight of 1919*. Naval Institute Press, Annapolis, Maryland, 1986.

Steiman, Hy and Kittler, Glenn. *Triumph*. Harper and Row, New York, 1961.

White, Ann Breese. *The First Transatlantic Flight: The triumph of the NC-4*. Talk given to the Denver Fortnightly Club, February 1985.

Wilbur, Commander Ted. *The First Flight Across the Atlantic, May 1919*. Smithsonian Institution, National Air and Space Museum, Washington, D.C., 1969.

Chapter 4

Chicago History Museum, Sterling Morton Papers

Chapter 5

Gray, Robert. "Oilheating Shangri La," *Fueloil & Oil Heat*, April 1951. (This trade journal is no longer published.)

Lowell, Steve. "James Breese: The Man Who Flew the Atlantic in the NC-4 is Now a New Mexico Industrialist," *Sun Trails* magazine, Volume 7, Number 8, October-November 1954, pp. 16-25. (This magazine is no longer published.)

Chapter 6

White, Ann Breese. *Recuerdos de los Vientos: Growing up in Santa Fe*. Talk given to the Denver Fortnightly Club, February 1993.

Chapter 7

Fortune, editors. *100 Stories of Business Success: Case Histories of American Enterprise*. Simon and Schuster, New York, 1954, p. 134.

Gray, Robert. "Oilheating Shangri La," *Fueloil & Oil Heat*, April 1951. (This trade journal is no longer published.)

Lowell, Steve. "James Breese: The Man Who Flew the Atlantic in the NC-4 is Now a New Mexico Industrialist," *Sun Trails* magazine, Volume 7, Number 8, October-

November 1954, pp. 16-25. (This magazine is no longer published.)

Chapter 8

"Hayters are Moving to New Mexico," *The Scarsdale Inquirer*, Scarsdale, NY, May 20, 1938.

Kilham, Larry. "As I Remember Him," Prize-winning short biography of James L. Breese III, Southwest Writers Association, Albuquerque, New Mexico, September 2019.

Chapter 9

"Breese Burners v. United States," The United States Court of Claims, Case number 50191, 1954-1957.

Chapter 10

Lowell, Steve. "James Breese: The Man Who Flew the Atlantic in the NC-4 is Now a New Mexico Industrialist," *Sun Trails* magazine, Volume 7, Number 8, October-November 1954, pp. 16-25. (This magazine is no longer published.)

Chapter 11

Murphey, John. "James L. Breese, Jr., Residence," Historical Cultural Properties Inventory Detail Form,

Historic Preservation Division, New Mexico Department of Cultural Affairs.

"Rob Wagner's *Script*, The Beverly Hills Magazine, 1929-1947," http://www.oldmagazinearticles.com/Rob_Wagner_Script_Magazine.

Chapter 12

McKinney, Robert, Editor and Publisher. "Jim Breese is Dead," *The New Mexican*, April 12, 1959, p. 4.